Enzymes in Carbohydrate Synthesis

A C S   S Y M P O S I U M   S E R I E S   **466**

Enzymes in Carbohydrate Synthesis

Mark D. Bednarski, EDITOR
University of California–Berkeley

Ethan S. Simon, EDITOR
Rohm and Haas

Developed from a symposium sponsored
by the Division of Carbohydrate Chemistry
at the 199th National Meeting
of the American Chemical Society,
Boston, Massachusetts,
April 22–27, 1990

American Chemical Society, Washington, DC 1991

TP
248
.65
E59
E592
1991
CHEM

Library of Congress Cataloging-in-Publication Data

Enzymes in carbohydrate synthesis /   Mark D. Bednarski, editor, Ethan S. Simon, editor.

    p.   cm.—(ACS symposium series, ISSN 0097–6156 ; 466)

"Developed from a symposium sponsored by the Division of Carbohydrate Chemistry at the 199th National Meeting."

Includes bibliographical references and index.

ISBN 0–8412–2097–2

1. Enzymes—Biotechnology—Congresses.   2. Carbohydrates—Biotechnology—Congresses. 3. Carbohydrates—Synthesis—Congresses.

I. Bednarski, Mark D., 1958–   . II. Simon, Ethan S., 1963–   . III. American Chemical Society. Division of Carbohydrate Chemistry. IV. Series.

TP248.65.E59E592   1991
660'.634—dc20   91–18662
                                           CIP

ACS Symposium Series

M. Joan Comstock, *Series Editor*

1991 ACS Books Advisory Board

Foreword

THE ACS SYMPOSIUM SERIES was founded in 1974 to provide a medium for publishing symposia quickly in book form. The format of the Series parallels that of the continuing ADVANCES IN CHEMISTRY SERIES except that, in order to save time, the papers are not typeset, but are reproduced as they are submitted by the authors in camera-ready form. Papers are reviewed under the supervision of the editors with the assistance of the Advisory Board and are selected to maintain the integrity of the symposia. Both reviews and reports of research are acceptable, because symposia may embrace both types of presentation. However, verbatim reproductions of previously published papers are not accepted.

Contents

APPENDIX

INDEXES

Preface

CARBOHYDRATES PLAY AN IMPORTANT ROLE in the interaction of cells with their environment. It is now well documented that glycoproteins and glycolipids mediate the attachment of pathogenic organisms to cells. These molecules are also involved in the processes of oncogenesis (the transformation of normal cells into cancer cells), metastasis (the spread of cancer through the body), and the targeting of leukocytes to areas of infection. Carbohydrate groups also influence protein structure and are involved in protein folding and in stabilizing proteins that come into contact with the external environment of the cell. Molecules that inhibit the process of assembling carbohydrates on proteins are being tested as drugs against viruses and bacteria and as mediators of the immune system.

The importance of glycoconjugates in biological processes has triggered a massive effort to synthesize complex carbohydrates. The total chemical synthesis of complex cell–surface carbohydrates such as gangliosides, globosides, and other complex glycolipids has recently been accomplished. However, despite these achievements, methods to selectively introduce key carbohydrates such as sialic acid and fucose into complex glycoproteins and glycolipids has remained an almost impossible task. This problem is further compounded by the increasing need to synthesize large amounts of complex oligosaccharides for medical trials and to change the structure of glycoproteins and glycolipids attached to recombinant proteins. Other desirable modifications may include the attachment and modification of oligosaccharides on proteins expressed in bacteria or yeast and the introduction of unnatural oligosaccharides onto lipids and proteins to increase their stability and enhance their biological activity.

The redesign of carbohydrate molecules attached to the surface of cells or to proteins and lipids cannot be accomplished using chemical methods. The use of enzymatic transformations, therefore, allows the development of technology that can be used directly in biological systems. Enzymatic reactions can also be used for large-scale production of complex monosaccharides, derivatives of monosaccharides, oligosaccharides, and polysaccharides. From an operational standpoint, the coupling of unprotected sugars to construct oligosaccharides would be far superior for the practical synthesis of these materials than the use of chemical methods employing elaborate protecting group strategies.

The chapters of this book introduce the reader to current results and state-of-the-art methods for the enzymatic synthesis of carbohydrates. They will give the reader an overall view of this field and what has been accomplished using commercially available enzymes. In addition, new enzymes that have recently been cloned and expressed are evaluated for their use in the synthesis of complex carbohydrates. We feel that the advances in molecular biology will have a profound influence on the enzymatic synthesis of carbohydrates. The first chapter (Toone and Whitesides) gives an overall view of the use of enzymes in the synthesis of monosaccharides and the use of activated-sugar nucleosides in oligosaccharide and polysaccharide synthesis. The second chapter (Wong) focuses on the construction of complex monosaccharides using aldolases. The third chapter (Hindsgaul et al.) discusses the use of glycosyltransferases in the synthesis of unnatural oligosaccharides. Chapters four (Nilsson) and five (Stangier and Thiem) focus on the use of glycosidases in conjunction with glycosyltransferases for the construction of complex oligosaccharides. Chapter six (Gygax et al.) demonstrates the use of glycosyltransferases for the modification of pharmaceutical agents; and chapter seven (Zehavi and Thiem) describes a method for the solid-phase oligosaccharide synthesis using enzymatic, coupling reactions. Chapter eight (Mazur) describes the use of galactose oxidase for the construction of unnatural galactosides; and chapter nine (Nakajima et al.) demonstrates the use of cofactor regeneration for the large-scale synthesis of sugar phosphates.

We have encouraged the authors to include details regarding the practical aspects of the enzymology involved in synthesis. Therefore, terms related to biochemical processes are explained in chemical terms and vice versa so either the biochemist or the organic chemist can feel comfortable with the material. Of course, it is always difficult to work at the interface between two fields, but we feel that this book has accomplished its goal if the reader becomes as excited as we are about the use of enzymes to solve problems in the synthesis of complex carbohydrates.

Acknowledgments

We wish to thank the symposium contributors for their efforts to help write this book. We would also like to thank the following companies for contributing to the symposium: Unitika Ltd., Monsanto Agricultural Company, Glaxo, Genzyme, Boehringer Mannheim Biochemicals, Rohm and Haas, Procter and Gamble, Polaroid Corporation, and the Division of Carbohydrate Chemistry of the American Chemical Society. A special thanks to Unitika Ltd. for their generous support. The peer reviewers deserve special recognition for their conscientious review of the chapters in this book. We are also indebted to Monique Funnié, University of California—Berkeley, who coordinated the preparation of the book for publication.

MARK D. BEDNARSKI
University of California
Berkeley, CA 94720

ETHAN S. SIMON
Rohm and Haas
Spring House, PA 19477

March 7, 1991

Chapter 1

Enzymes as Catalysts in Carbohydrate Synthesis

Eric J. Toone[1] and George M. Whitesides

Department of Chemistry, Harvard University, Cambridge, MA 02138

The discovery of the myriad roles of carbohydrates in biological recognition phenomena has led to an increased demand for practical routes to gram-scale quantities of unnatural carbohydrate structures. We have investigated the applicability of enzymes as catalysts for mono- oligo-, and polysaccharide synthesis. Here, we review our progress in the area, in particular the use of aldolases for monosaccharide synthesis, the synthesis of the activated sugar nucleosides required for Leloir-pathway glycosidic bond formation, and the use of both Leloir- and non-Leloir-pathway enzymes for the synthesis of carbohydrate polymers.

The past two decades have seen explosive growth in the field of carbohydrate chemistry. The synthesis and structural analysis of natural and unnatural carbohydrate structures has reached levels impossible only a few years ago. To a large extent, growth in carbohydrate chemistry has been driven by advances in carbohydrate biology, or glycobiology. Until recently, carbohydrates were generally regarded solely as energy storage vehicles and structural units in cells. These notions have now been challenged: it is clear that carbohydrates govern a wide range of biological recognition phenomena (1-3). In addition to their well-known recognition roles as blood group antigens, it is now known that carbohydrates act as binding sites for a wide range of bacteria, viruses, hormones and soluble toxins (4-7). Cell-surface carbohydrates also play key roles in intercellular communication events that control growth and differentiation as well as organogenesis (8). In addition to these now well-defined roles, research is continually establishing new roles for carbohydrates in other systems, including, for instance, as binding sites for cell-adhesion molecules (CAMs) (9) and tumor necrosis factor (10).

With these new-found roles for carbohydrates has come an increased demand for practical synthetic routes to this class of compounds. An enormous number of carbohydrate and carbohydrate-like structures exist that are of biomedical interest (Figure 1). Despite efforts towards general methodologies, carbohydrates remain one of the most challenging groups of compounds to prepare (11-14). In the past, enzymes have been

[1]Current address: Department of Chemistry, Duke University, Durham, NC 27706

0097–6156/91/0466–0001$06.50/0

utilized as catalysts for the synthesis of a range of non-carbohydrate compounds (15-17). More recently, enzymes have been applied to carbohydrate synthesis (18). Briefly, enzymes offer two major advantages over classical methodologies for the synthesis of carbohydrates:

i) *Compatibility with aqueous media.* Many of the reactions commonly utilized by organic chemists are incompatible with water. Water, however, is the most practical medium for synthetic manipulations of unprotected, hydrophilic compounds such as carbohydrates. Enzymes generally operate in aqueous solution, at or near neutral pH, at or near room temperature. The use of enzymes as catalysts for carbohydrate synthesis therefore avoids the necessity for protection/deprotection schemes.

ii) *Specificity.* The specificity of enzymes is manifested in three major ways, all of which are vital to successful carbohydrate syntheses. First, enzyme-catalyzed reactions demonstrate absolute *chemospecificity.* A second important feature of enzyme-catalyzed reactions is high *regiospecificity.* Since carbohydrates generally contain a number of hydroxyl groups of approximately equal reactivity, the ability to selectively manipulate a single hydroxyl residue is clearly important. Third, enzymes display exquisite *stereospecificity.* Carbohydrates are chirotopic species, and usually possess multiple stereogenic centers. This stereochemical information must be correctly installed in any successful synthetic methodology.

It is clear, then, that perhaps no other group of compounds is as well suited to enzyme-based synthesis as is the carbohydrates. During the past several years, we have developed a range of enzyme-based syntheses for mono-, oligo-, and polysaccharides. This paper reviews some of our efforts on the use of aldolases and transketolase for monosaccharide synthesis, as well as the use of Leloir and non-Leloir pathway enzymes for glycosidic bond formation.

Monosaccharide Synthesis Using Aldolases

FDP Aldolase. The most extensively utilized class of enzymes for monosaccharide synthesis are the aldolases (E.C. sub-class 4.1.2.). This ubiquitous group of enzymes catalyzes reversible aldol reactions *in vivo.* Two major groups of aldolases exist: type I aldolases, found primarily in higher plants and animals, catalyze aldol condensations by means of a Schiff base formed between an enzyme lysine e-amino group and the nucleophilic carbonyl group; type II aldolases, found primarily in microorganisms, utilize a divalent zinc to activate the nucleophilic component (19). Approximately 25 aldolases have been identified to date (18).

Aldolases have been studied as catalysts for monosaccharide synthesis for nearly 40 years. The best studied member of the group is a fructose-1,6-diphosphate (FDP) aldolase from rabbit muscle (RAMA, E.C. 4.1.2.13) (20). *In vivo*, this enzyme catalyzes the reversible condensation of D-glyceraldehyde-3-phosphate and dihydroxyacetone phosphate (DHAP) to generate FDP (Scheme 1). In the synthetic direction, the enzyme catalyzes the formation of two new stereogenic centers with absolute stereospecificity: the stereochemistry of the new vicinal diol is always D-*threo.* RAMA will accept a wide range of aldehydes as electrophiles in the aldol condensation. Studies have demonstrated that virtually any aldehyde except aldehydes sterically hindered at the α-position, α,β-unsaturated

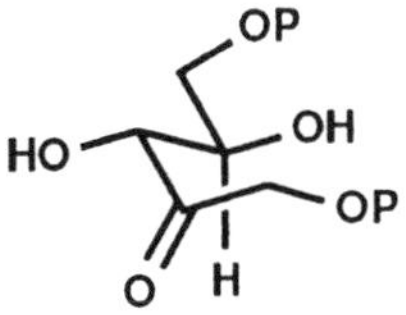

Ribulose-1,5-P$_2$ 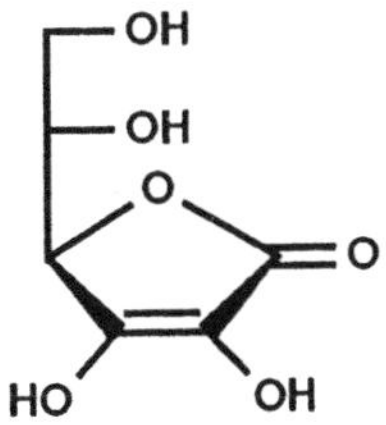   Ascorbic Acid

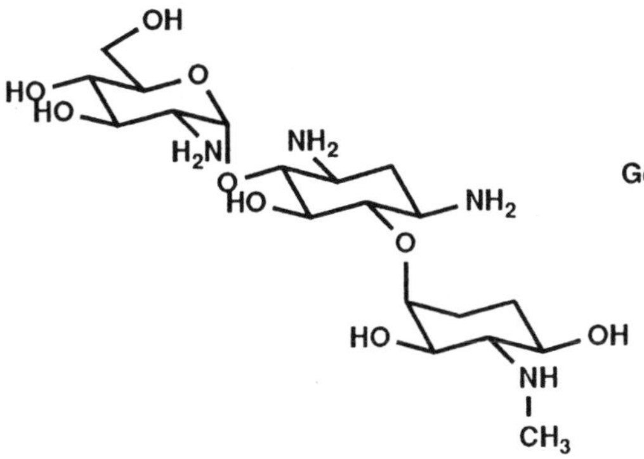

Gentamycin A

Figure 1.   Carbohydrate Structures of Biomedical Interest.

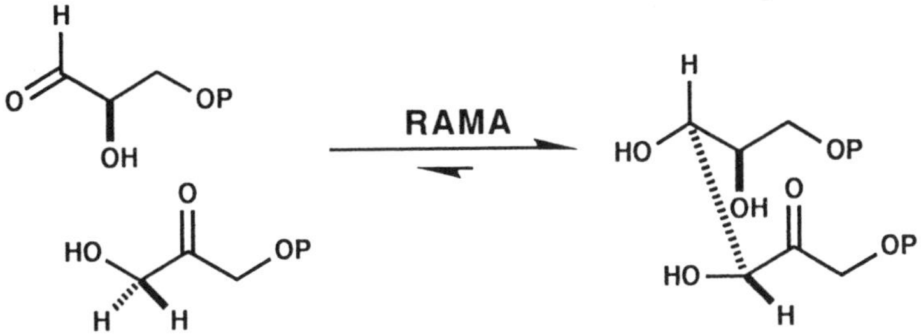

RAMA

Scheme 1.   The RAMA-Catalyzed Aldol Condensation.

aldehydes, or those that can readily eliminate to α,β-unsaturated aldehydes, is accepted as a substrate. The demand for DHAP as nucleophile seems to be absolute (18). We have prepared a number of useful monosaccharides using RAMA, including 3-deoxy-D-*arabino*-heptulosonic acid 7-phosphonate, an inhibitor of the shikimate pathway (DAHP, Scheme 2) (21) and Furaneol®, a flavoring component (22).

Despite the demonstrated utility of RAMA in carbohydrate synthesis, an inability to install stereochemistries other than D-*threo* at the C3/C4 vicinal diol limits the range of applications of the enzyme. We have pursued strategies designed to overcome this limitation. The "inversion strategy" (Scheme 3) makes use of a monoprotected dialdehyde as the electrophile during a RAMA-catalyzed condensation (23). The resulting ketose can then be reduced, and the protected aldehyde functionality unmasked to generate a new aldose sugar.

The stereospecific reduction of the ketone functionality is clearly of paramount importance to the inversion strategy. We have used iditol dehydrogenase (also known as sorbitol dehydrogenase or polyol dehydrogenase, E.C. 1.1.1.14) from *Candida utilis* or sheep liver to selectively generate *either* diastereomer of the new polyol using a single oxido-reductase (23). Reduction of the ketose with iditol dehydrogenase generates the 2R polyol exclusively (Scheme 4). Alternatively, chemical reduction of the ketone followed by stereospecific *oxidation* of the unwanted enantiomer with iditol dehydrogenase generates the 2S diastereomer (Scheme 5). The ketose resulting from the unwanted enantiomer can then be recycled.

Fuculose 1-Phosphate Aldolase. Another potential route to different stereochemistries at C3 and C4 is *via* different aldolases. Of the four possible diastereomers that can result from an aldol condensation between DHAP and an electrophile, aldolases have been identified in Nature that stereoselectively generate three (Scheme 6).

We have recently overexpressed a fuculose 1-phosphate (Fuc-1-P) aldolase from *E. coli* and have begun to evaluate its utility as a catalyst for carbohydrate synthesis (24). We have expressed the enzyme using a *tac* promoter, to a level of 6×10^3 units per liter (one unit of enzyme catalyzes the conversion of 1 μmol of substrates to products per minute under optimal conditions of temperature and pH). The enzyme is easily purified to crystallinity using ion-exchange chromatography. Fuc-1-P aldolase has a usefully broad substrate specificity, at least with regard to the electrophilic component: the enzyme appears to accept over 40 compounds in kinetic assays. Investigations on the substrate specificity with regard to the nucleophilic component remain to be completed. Fuc-1-P aldolase has been used in a preparative-scale (10 μmol) synthesis of D-ribulose (Scheme 7) (24).

Synthesis of DHAP. Both RAMA and Fuc-1-P aldolase require DHAP as the nucleophilic component. Although DHAP is commercially available, it is too expensive for synthetic use, and must be synthesized for preparative-scale enzymatic reactions. DHAP has been prepared *via* three major routes: enzymatically from FDP using a combination of RAMA and triose isomerase (TIM, E.C. 5.3.1.1) (25), chemically, by phosphorylation of dihydroxyacetone dimer (26), and enzymatically by phosphorylation of dihydroxyacetone catalyzed by glycerokinase (Scheme 8) (25).

The preparation of DHAP by phosphorylation of dihydroxyacetone with glycerokinase is most effective for large-scale (mole) syntheses of

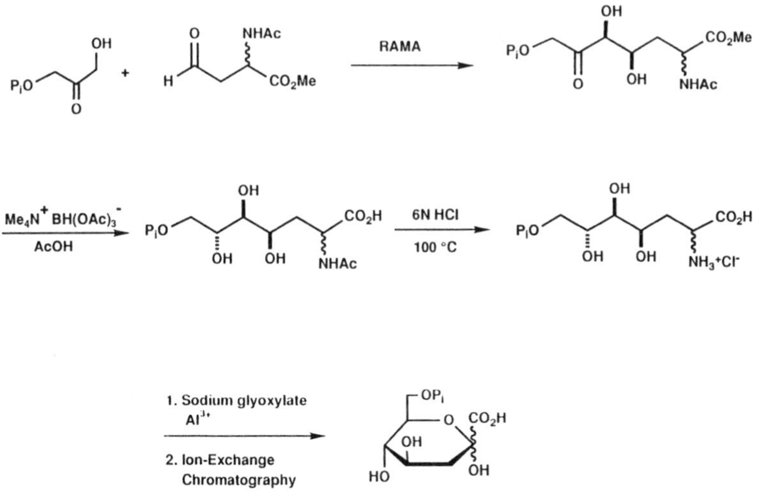

Scheme 2.   Synthesis of DAHP.

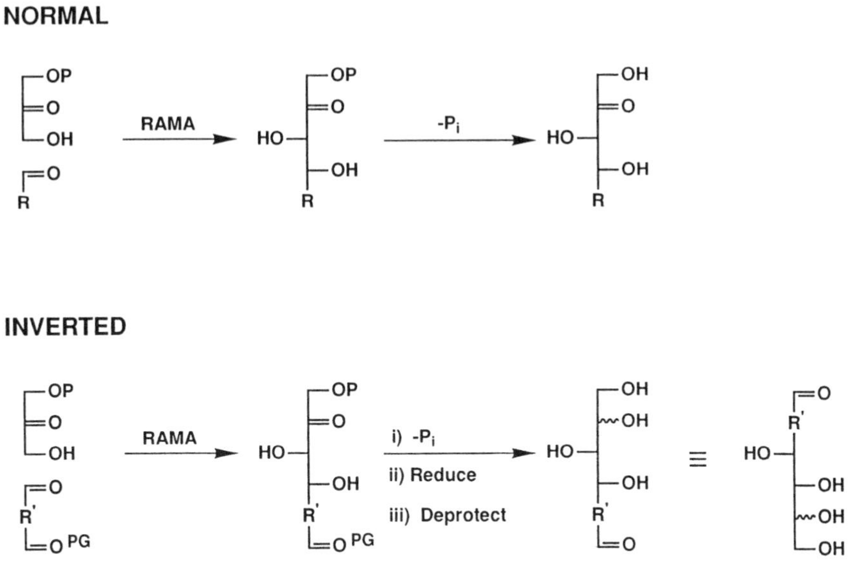

NORMAL

INVERTED

Scheme 3.   The Inversion Strategy.   P = phosphate; PG = protecting group.

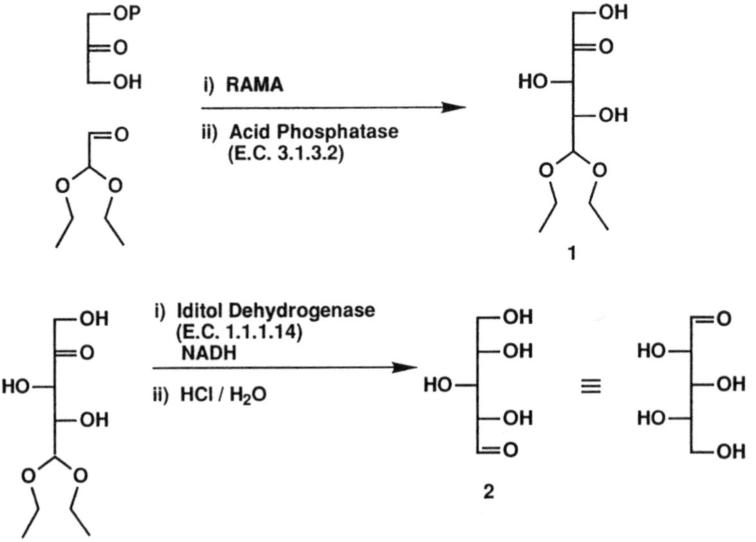

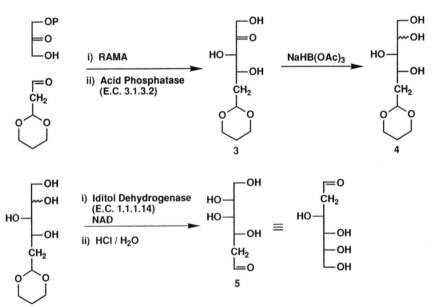

Scheme 4.   Synthesis of L-Xylose.

Scheme 5.   Synthesis of   2-Deoxy-D-*arabino*-hexose.

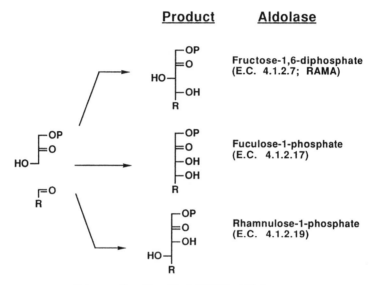

Scheme 6.   Identified DHAP Aldolases.

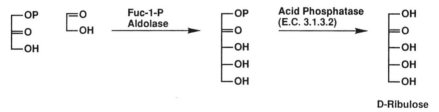

Scheme 7.   Synthesis of D-Ribulose.

DHAP (27). This strategy, however, requires an effective regeneration scheme for ATP. We have utilized two regeneration schemes for ATP, one based on pyruvate kinase (E.C. 2.7.1.40), and one based on acetyl kinase (E.C. 2.7.2.1). Both phosphate donors, acetyl phosphate and phosphoenolpyruvate (PEP) must be prepared: both have been prepared chemically on a mole scale (27).

Recently, we have developed a method for preparing PEP *in situ* from commercially available 3-phosphoglyceric acid (PGA), using phosphoglyceromutase and enolase (Scheme 9) (28).

Other Aldolases. In addition to the DHAP aldolases, we have conducted preliminary investigations of two other aldolases. KDO synthase (E.C. 4.1.2.16) catalyzes the formation of 2-keto-3-deoxy-L-*arabino*-octulosonic acid 8-phosphate (KDO-8-P) from arabinose 5-phosphate and PEP (Scheme 10) (29). KDO is an integral component of Gram-negative bacterial cell walls, and derivatives of KDO are of interest as inhibitors of cell wall formation (30, 31).

N-Acetylneuraminic acid (NeuAc) aldolase (E.C. 4.1.3.3) catalyzes the formation of NeuAc from *N*-acetylmannosamine and pyruvate (Scheme 11). The sialic acids are key cell-surface determinants of mammalian glycoconjugates. We and others have prepared *N*-acetylneuraminic acid (32) *via* a NeuAc aldolase-catalyzed condensation between *N*-acetylmannosamine and pyruvate. A number of derivatives of sialic acid have been prepared using derivatives of mannose (33-35).

Transketolase. Another group of enzymes that catalyze the stereospecific formation and cleavage of carbohydrates *in vivo* are the transketolases and transaldolases. Transketolase (E.C. 2.2.1.1) is a thiamin pyrophosphate (TPP) dependent enzyme that catalyzes the transfer of a hydroxyketo group from a ketose phosphate to an aldose phosphate in the pentose pathway (Scheme 12) (36).

The action of transketolase generates vicinal diols having the same stereochemistry as the products of RAMA-catalyzed condensation. The enzyme, however, has two significant advantages over RAMA: the reaction does not require DHAP, and the products are not phosphorylated. The ketose functionality can be replaced by hydroxy pyruvate, which provides a hydroxyketo equivalent after decarboxylation. No other hydroxy acid has yet been found that is accepted by transketolase. Although the enzyme is absolute in its requirement for the R configuration of the hydroxy functionality at C2 of the aldehyde, there seem to be no other stereochemical requirements. Transketolase accepts a range of aldoses as substrates, and should be a useful enzyme for carbohydrate synthesis (Table 1) (37).

Synthesis of Oligo- and Polysaccharides Using Glycosyl-transferases

One of the greatest single problems in the field of carbohydrate synthesis is the development of reliable methods for the formation of glycosidic bonds. Despite the development of numerous protocols for the synthesis of oligosaccharides, the available coupling reactions often give mixtures of anomers, occur with low yields, and lack generality (38). The formation of glycosidic linkages is another area of carbohydrate chemistry where enzymes are now beginning to have an impact.

The glycosyltransferases transfer activated monosaccharides to nascent oligo- or polysaccharide chains. There are two basic motifs by

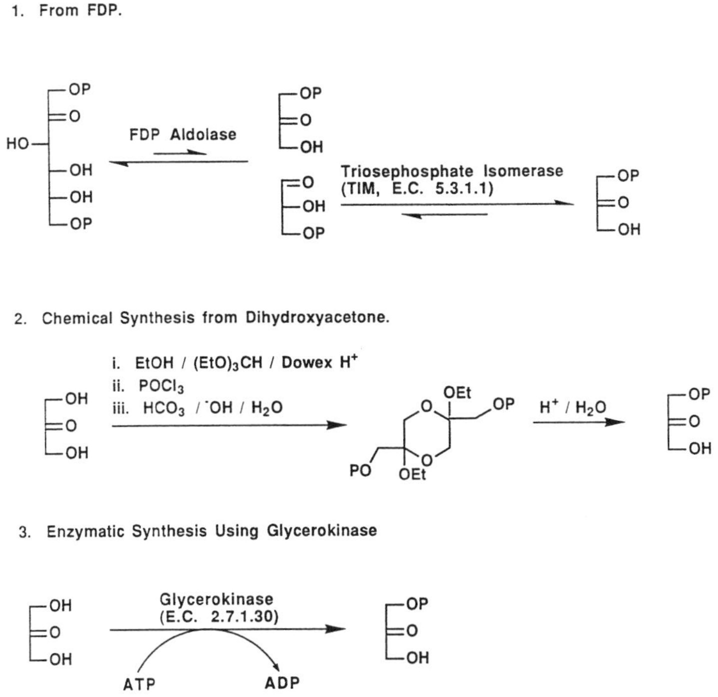
1. From FDP.
FDP Aldolase
HO
OP
O
OH
Triosephosphate Isomerase
(TIM, E.C. 5.3.1.1)
2. Chemical Synthesis from Dihydroxyacetone.
i. EtOH / (EtO)3CH / Dowex H+
ii. POCl3
iii. HCO3 / -OH / H2O
OEt
PO
H+ / H2O
3. Enzymatic Synthesis Using Glycerokinase
Glycerokinase
(E.C. 2.7.1.30)
ATP
ADP

Scheme 8. Synthesis of DHAP.

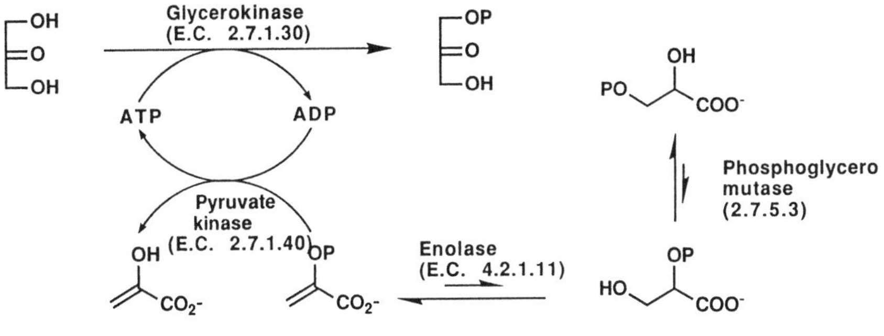
Glycerokinase
(E.C. 2.7.1.30)
ATP
ADP
Pyruvate
kinase
(E.C. 2.7.1.40)
OH
OP
CO2-
Enolase
(E.C. 4.2.1.11)
PO
HO
COO-
Phosphoglycero
mutase
(2.7.5.3)

Scheme 9. Enzymatic Synthesis of DHAP.

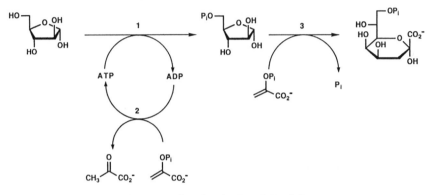

Scheme 10.   Synthesis of KDO-8-P.

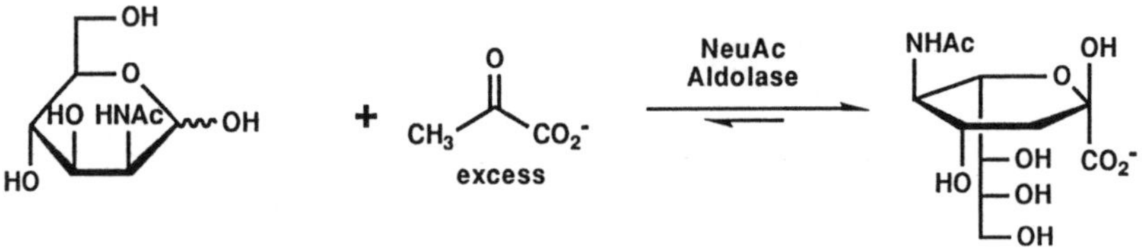

Scheme 11.   Synthesis of NeuAc.

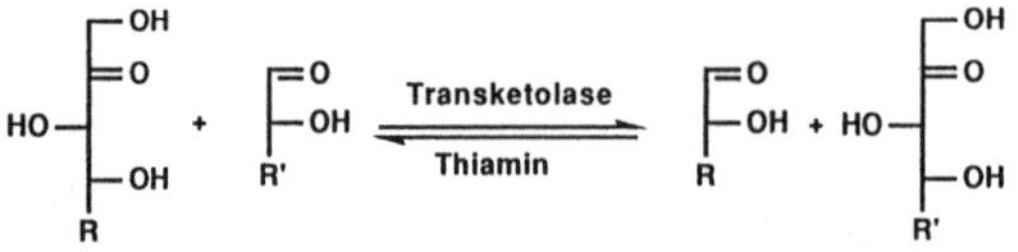

Scheme 12.   Transketolase-Catalyzed Ketose Transfer.

Table 1.   Substrate Specificity of Transketolase

Aldehyde Substrates:

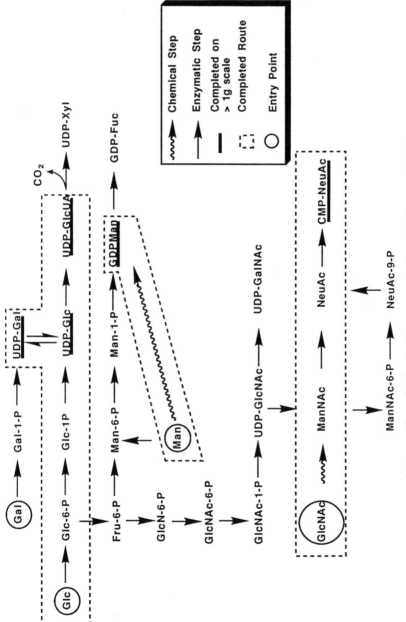

Scheme 13.   Synthetic Routes to Activated Monosaccharides.

which monosaccharides are activated *in vivo*. The majority of glycosyltransferases utilize sugars activated as the nucleoside phosphate derivatives. The enzymes that catalyze the formation and transfer of nucleoside phosphate sugars are collectively referred to as the enzymes of the Leloir pathway, after the Argentinian biochemist who elucidated this biosynthetic route. A second group of glycosyltransferases use monosaccharides activated as the sugar-1-phosphate. Examples of this class of enzymes are sucrose phosphorylase and amylose phosphorylase. We have conducted investigations over the past several years on both groups of enzymes as catalysts for glycosidic bond formation.

   Leloir Pathway Glycosyltransferases. The Preparation of Activated Monosaccharides. In mammalian biochemistry, eight monosaccharides are commonly encountered: uridine 5'-diphosphoglucose (UDP-Glc), uridine 5'-diphosphogalactose (UDP-Gal), uridine 5'-diphosphoglucuronic acid (UDP-GlcUA), uridine 5'-diphospho-*N*-acetyl-glucosamine (UDP-GlcNAc), uridine 5'-diphospho-*N*-acetylgalactose (UDP-GalNAc), guanidine 5'-diphosphomannose (GDP-Man), guanidine 5'-diphosphofucose (GDP-Fuc), and cytidine 5'-monophospho-*N*-acetyl-neuraminic acid (CMP-NeuAc). A substantial portion of our research in the area of Leloir-pathway synthesis has focussed on the development of practical synthetic routes to multigram quantities of the nucleoside phosphate sugars required to transfer these eight sugars. The biosynthetic routes by which these sugars are synthesized are shown in Scheme 13. Our progress in this area is also shown schematically. We discuss below our recent efforts towards the syntheses of UDP-GlcUA, and CMP-NeuAc. We have previously reported multigram preparations of UDP-Glc and UDP-Gal (*39*).

   Synthesis of Nucleoside Triphosphates (XTPs). *In vivo*, the nucleoside diphosphate sugars are synthesized from the sugar-1-phosphate and the appropriate nucleoside triphosphates (Scheme 14).

$$\text{Sugar-1-P} + \text{XTP} \rightarrow \text{XDP-Sugar}$$

Scheme 14.   Biosynthesis of Nucleoside Diphosphate Sugars.

An exception is the synthesis of CMPNeuAc, which proceeds directly from neuraminic acid (Scheme 15).

$$\text{NeuAc} + \text{CTP} \rightarrow \text{CMP-NeuAc}$$

Scheme 15.   Biosynthesis of CMPNeuAc.

   Clearly, good synthetic routes to all of the required XTPs is a prerequisite to successful preparation of activated nucleoside phosphate sugars. Although literature methods exist for the preparation of all XTPs from the corresponding XMP, few methods for the convenient preparation of gram quantities exist. We therefore undertook an examination of the various potential routes, both enzymatic and chemical, to determine the optimal preparation of each XTP. Scheme 16 outlines the most effective enzymatic route to the nucleoside triphosphates (*40*).

   Adenylate kinase, which *in vivo* catalyzes the equilibrium between adenosine mono-, di- and triphosphates, has been used extensively in the production of ATP (*27, 41*). Although the enzyme has a broad substrate specificity for nucleoside di- and triphosphates, the specificity for monophosphates is much more restrictive. Nonetheless, the specificity is broad enough to permit a multigram synthesis of CTP (*42*). Guanylate kinase offers the most effective route to GTP, while UTP is best prepared by chemical deamination of CTP (*40*). In all cases the ultimate phosphate donor is PEP, which can be generated from PGA (*vide supra*). All of the nucleoside monophosphates are commercially available at low cost.

UDP-GlcUA. Glucuronic acid occurs *in vivo* as a conjugate for xenobiotic removal (*43*). The uronic acids (glucuronic acid and L-iduronic acid) also occur in the glycoamineglycan polysaccharides, previously referred to as the mucopolysaccharides. Important members of the glycosamineglycans include hyaluronic acid, chondroitin, and heparin. Iduronic acid is not transferred as a monosaccharide: it is generated by epimerization of glucuronic acid in an intact polymer.

UDP-GlcUA is synthesized *in vivo* by oxidation of UDP-Glc. The nicotinamide dependent UDP-Glc dehydrogenase (E.C. 1.1.1.22, from bovine liver) is commercially available at $20/U. In our hands, the enzyme was unstable, and unsuitable for the preparation of gram-scale quantities of UDP-GlcUA. UDP-Glc dehydrogenase can be isolated from whole bovine liver, according to literature methods (*44*). This isolation yields approximately 450 U of enzyme from 2.5 kg of frozen liver. We were able to prepare UDP-GlcUA from UDP-Glc on a 1-gram scale using this preparation (Scheme 17). The overall yield was 76% (91% mass recovery, 84% pure by enzymatic assay) (*45*). The nicotinamide cofactor was regenerated using pyruvate with L-lactate dehydrogenase.

We were able to synthesize the simplest member of the glycosamine-glycans, hyaluronic acid, using a cell-free system of enzymes isolated from *Streptococcus zooepidemicus* (Scheme 18) (*46*). UDP-GlcUA was generated *in situ* from UDP-Glc using commercially available UDP-Glc dehydrogenase. Although the current enzyme preparation is not suitable for large-scale production of hyaluronic acid, the cell-free enzyme system offers the possibility of incorporating unnatural carbohydrates into the polymer. This could potentially lead to the synthesis of polymers with more desirable properties than the naturally occurring material.

CMP-NeuAc. Derivatives of neuraminic acid frequently terminate mammalian glycoconjugates. Activated NeuAc is therefore an especially important target for enzyme-based synthesis. NeuAc is synthesized *in vivo* by an NeuAc aldolase-catalyzed condensation of *N*-acetylmannosamine and pyruvate (Scheme 11 above). NeuAc is then coupled directly to CTP by CMP-NeuAc synthase (E.C. 2.7.7.43). We have recently published a multigram preparation of CMP-NeuAc which makes use of both NeuAc aldolase and CMP-NeuAc synthase (Scheme 19) (*47*). NeuAc aldolase has been cloned and overexpressed, and is commercially available. The final enzyme required, CMP-NeuAc synthase, was isolated from calf brain.

Non-Leloir Pathway Glycosyltransferases. Glycosidic linkages have also been formed using glycosyl transferases which utilize sugar-1-phosphates as activated monosaccharides. Both sucrose phosphorylase and trehalose phosphorylase have been utilized *in vitro* to synthesize disaccharides (Scheme 20) (*48*). Synthetic methodologies based on isolated enzymes as catalysts may allow the preparation of unnatural analogues of these two important sugars.

A number of polysaccharides can also be prepared using non-Leloir glycosyl transferases. Maltooligomers could be polymerized by potato phosphorylase (PPh, E.C. 2.4.1.1). The activated glucose 1-phosphate in this scheme was generated *in situ* from sucrose and inorganic phosphate by the action of sucrose phosphorylase (SPh, E.C. 2.4.1.7, Scheme 21) (*49*).

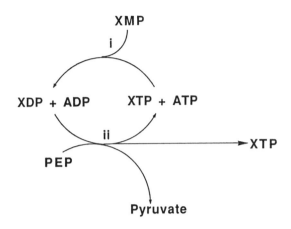

i) **Adenylate Kinase**
    (E.C. 2.7.4.3, X = A, C, U)

Guanylate Kinase
(E.C. 2.7.4.8, X = G)

Nucleoside Monophosphate Kinase
(E.C. 2.7.4.4, X = U)

ii) **Pyruvate Kinase**
     (E.C. 2.7.1.40)

Scheme 16.   Enzymatic Syntheses of Nucleoside Triphosphates.

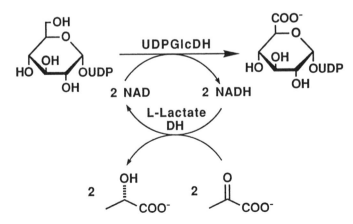

Scheme 17.   Enzymatic Synthesis of UDP-GlcUA.

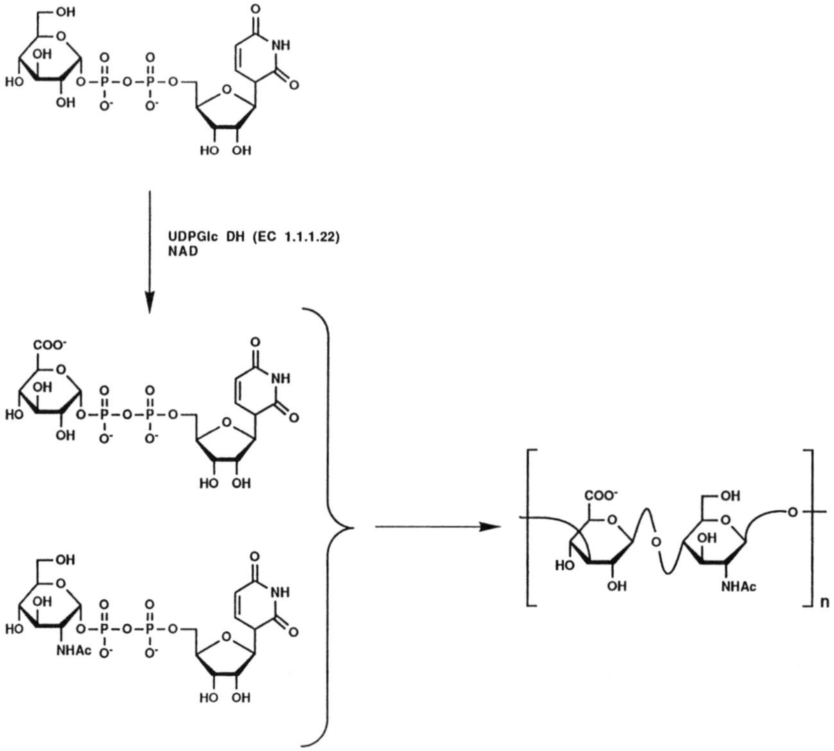

Scheme 18.   Cell-Free Synthesis of Hyaluronic Acid.

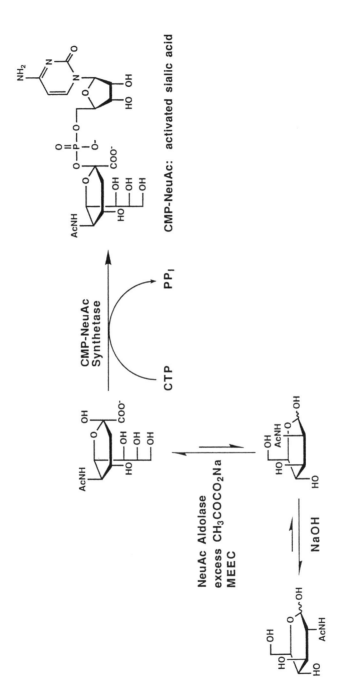

Scheme 19. Enzymatic Synthesis of CMP-NeuAc.

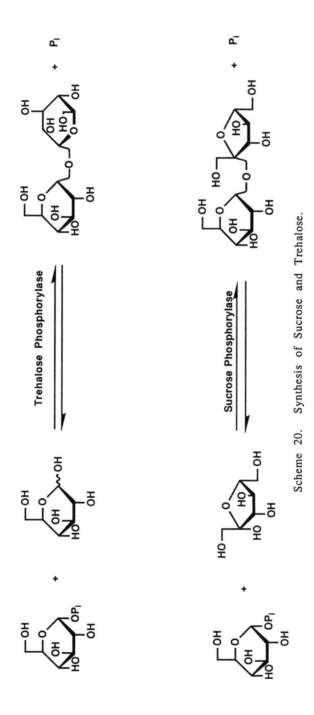

Scheme 20.   Synthesis of Sucrose and Trehalose.

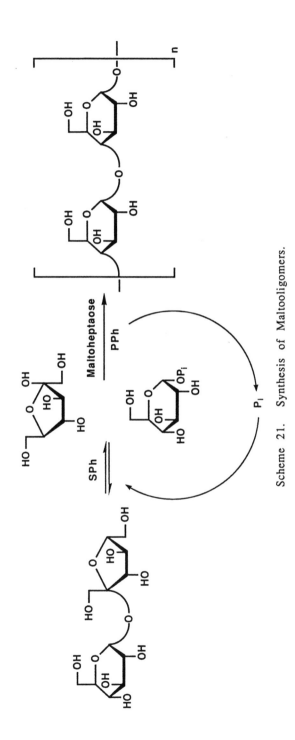

Scheme 21.   Synthesis of Maltooligomers.

Pfannemüller and coworkers have used similar methodologies to prepare an interesting class of linear, star and comb-shaped polymers (50, 51).

This strategy was employed to add amylose polymers to RNAse (Scheme 22) (49). It is clear that the carbohydrate component of glycoproteins enhances the stability *in vivo* of circulating proteins, and recombinant proteins modified in this way may be useful therapeutic agents.

Conclusions

To meet the requirements for synthetic routes to a wide range of carbo-hydrates, we have begun investigating enzyme-based methodologies. The initial results demonstrate conclusively that enzyme technology will play an important role in the future. Although the field is currently limited to some extent by a lack of availability of enzymes, modern molecular biology will almost certainly overcome this problem in the foreseeable future.

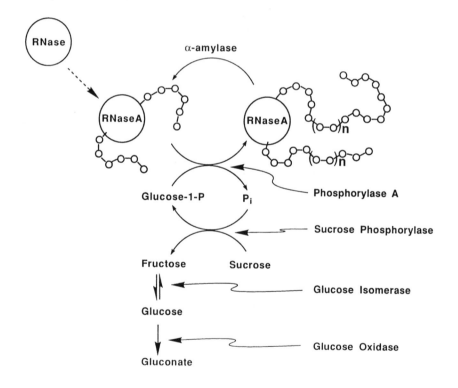

Scheme 22.   Growth of Amylose Polymers on RNAse.

Acknowledgments

This work was carried out by many members of the Whitesides group, past and present. The work presented here was supported by NIH grants GM 30367 and GM 39589.

Literature   Cited

1.   Sharon, N. *Complex Carbohydrates*; Addison-Wesley: Reading, MA, 1975.
2.   a) *The Glycocongugates, Vol I - II*; Horowitz, M. I.; Pigman, W., Eds., Academic: New York, 1977-1978. b) *The Glycoconjugates, Vol III - IV*; Horowitz, M. I., Ed., Academic: New York, 1982.
3.   Collins, P. M. *Carbohydrates*; Chapman and Hall: London, 1987.
4.   Schauer, R. *Adv. Carbohydr. Chem. Biochem.* **1982**, *40*, 131.
5.   *The Lectins: Properties, Functions and Applications in Biology and Medicine*; Liener, I. E.; Sharon, N.; Goldstein, I. J., Eds.; Academic: New York, 1986.
6.   Paulson, J. C. In *The Receptors*; Cohn, P. M., Ed.; Academic: New York, 1985; Vol. 2, p 131.
7.   Sairam, M. R. In *The Receptors*; Cohn, P. M., Ed.; Academic: New York, 1985; Vol. 2, p 307.
8.   Rademacher, T. W.; Parekh, R. B.; Dwek, R. A. *Ann. Rev. Biochem.* **1988**, *57*, 785.
9.   Rice, G. E.; Bevilacqua, M. P. *Science* **1989**, *246*, 1303.
10. Smith, C. A.; Davis, T.; Anderson, D.; Solam, L.; Beckman, M. P.; Jerzy, R.; Dower, S. K.; Cosman, D.; Goodwin, R. G. *Science* **1990**, *248*, 1019.
11. Schmidt, R. R. *Angew. Chem. Int. Ed. Engl.* **1986**, *25*, 212.
12   Kunz, H. *Angew. Chem. Int. Ed. Engl.* **1987**, *26*, 294.
13. Paulsen, H. *Angew. Chem. Int. Ed. Engl.* **1982**, *21*, 155.
14. Lemieux, R. U. *Chem. Soc. Rev.* **1978**, *7*, 423.
15. Wong, C. -H. *Science* **1989**, *244*, 1145.
16. Jones, J. B. *Tetrahedron* **1986**, *42*, 3351.
17. *Enzymes in Organic Synthesis (Ciba Foundation Symposium 111)*; Porter, R.; Clark, S., Eds.; Pitman: London, 1985.
18. Toone, E. J.; Simon, E. S.; Bednarski, M. D.; Whitesides, G. M. *Tetrahedron* **1989**, *45*, 5365.
19. Horecker, L.; Tsolas, O.; Lai, C. Y. In *The Enzymes,* Boyer, P. D., Ed.; Academic: New York, 1972; Vol. 7, p 213.
20. Bednarski, M. D.; Simon, E. S.; Bischofberger, N.; Fessner, W. -D.; Kim, M. -J.; Lees, W.; Saito, T.; Waldmann, H.; Whitesides, G. M. *J. Am. Chem. Soc.* **1989**, *111*, 627.
21. Turner, N. J.; Whitesides, G. M. *J. Am. Chem. Soc.* **1989**, *111*, 624.
22. Wong, C. -H.; Mazenod, F. P.; Whitesides, G. M. *J. Org. Chem.* **1983**, *48*, 3493.
23. Borysenko, C.; Spaltenstein, A.; Straub, J.; Whitesides, G. M. *J. Am. Chem. Soc.* **1989**, *111*, 9275.
24. Ozaki, A.; Toone, E. J.; von der Osten, C. H.; Sinskey, A. J.; Whitesides, G. M. *J. Am. Chem. Soc.* **1990**, *112*, 4970.
25. Bednarski, M. D.; Waldmann, H. J.; Whitesides, G. M. *Tetrahedron Lett.* **1986**, *27*, 5807.
26. Effenberger, F.; Straub, A. *Tetrahedron Lett.* **1987**, *28*, 1641.
27. Crans, D. C.; Kazlauskas, R. J.; Hirschbein, B. L.; Wong, C. -H.; Abril, O.; Whitesides, G. M. *Methods Enzymol.* **1987**, *136*, 263.

28. Simon, E. S.; Grabowski, S.; Whitesides, G. M. *J. Am. Chem. Soc.*, **1989**, *111*, 8920.
29. Bednarski, M. D.; Crans, D. C.; DiCosimo, R.; Simon. E. S.; Stein, P. D.; Whitesides, G. M.;   Schneider, M. *Tetrahedron Lett.* **1988**, *29*, 427.
30. Hammond, S. M.; Claesson, A.; Jansson, A. M.; Larsson, L. -G.; Pring, B. G.; Town, C. M.; Ekström, B. *Nature* **1987**, *327*, 730.
31. Danishefsky, S. J.; DeNinno, M. P. *Angew. Chem. Int. Ed. Engl.* **1987**, *26*, 15.
32. Bednarski, M. D.; Chenault, H. K.; Simon, E. S.; Whitesides, G. M. *J. Am. Chem. Soc.* **1987**, *109*, 1283.
33. David, S.; Augé, C. *Pure Appl. Chem.* **1987**, *59*, 1501.
34. Kim, M. -J.; Hennen, W. J.; Sweers, H. M.; Wong, C. -H. *J. Am. Chem. Soc.* **1988**, *110*, 6481.
35. Augé, C.; Gautheron, C. *J. Chem. Soc. Chem. Commun.* **1987**, 859.
36. Racker, E. In *The Enzymes*; Boyer, P. D.; Lardy, H.; Myrbäch, K., Eds.; Academic: New York; 1961; Vol. 5, p 397.
37. Myles, D.; Kobori, Y.; Whitesides, G. M., unpublished results.
38. Paulsen, H. *Chem. Soc. Rev.* **1984**, *13*, 15.
39. Wong, C. -H.; Haynie, S. L.; Whitesides, G. M. *J. Org. Chem.* **1982**, *47*, 5416.
40. Simon, E. S.; Grabowski, S.; Whitesides, G. M. *J. Org. Chem.* **1990**, *55*, 1834.
41. Hirschein, B. ; Mazenod, F. P.; Whitesides, G. M. *J. Org. Chem.* **1982**, *47*, 3765.
42. Simon, E. S.; Bednarski, M. D.; Whitesides, G. M. *Tetrahedron Lett.* **1988**, *29*, 1123.
43. *Enzymatic Bases of Detoxication*; Jakoby, W. B., Ed.; Academic: New York, 1980; Vol. 2.
44. Zalitis, J.; Feingold, D. S. *Arch. Biochem. Biophys.* **1969**, *12*, 457.
45. Toone, E. J.; Simon, E. S.; Whitesides, G. M., unpublished results.
46. Simon, E. S.; Toone, E. J.; Ostroff, G.; Bednarski, M. D.; Whitesides, G. M. *Methods Enzymol.* **1989**, *179*, 275.
47. Simon. E. S.; Bednarski, M. D.; Whitesides, G. M. *J. Am. Chem. Soc.* **1988**, *110*, 7159.
48. Haynie, S. L., Whitesides, G. M. *Appl. Biochem. Biotech.* **1990**, *23*, 155.
49. Waldmann, H.; Gygax, D.; Bednarski, M. D.; Shangraw, W. R.; Whitesides, G. M. *Carbohydr. Res.* **1986**, *157*, c4.
50. Ziegast, G.; Pfannemüller, B. *Carbohydr. Res.* **1987**, *160*, 185, and references therein.
51. Ziegast, G.; Pfannemüller, B. *Makromol. Chem., Rapid Commun.* **1984**, *5*, 373.

RECEIVED April 4, 1991

Chapter 2

Microbial Aldolases in Carbohydrate Synthesis

C.-H. Wong

Department of Chemistry, Research Institute of Scripps Clinic,
10666 North Torrey Pines Road, La Jolla, CA 92037

This chapter describes recent developments in carbohydrate synthesis using microbial aldolases. Recombinant type II fructose-1,6-diphosphate aldolase and type I 2-deoxyribose-5-phosphate aldolase have been exploited for the synthesis of monosaccharides and analogs. A new procedure for the synthesis of dihydroxyacetone phosphate has been developed. Sialic acid aldolase has been used in conjunction with other enzyme-catalyzed modifications of monosaccharides for the synthesis of sialic acid-related sugars.

Catalytic asymmetric aldol condensations remain one of the most interesting and challenging subjects in synthetic organic chemistry (1-4). The enzymes which catalyze aldol reactions hold a great potential in this regard (5-7). Of more than 20 aldolases reported, eight have been exploited for organic synthesis. Two common features are often found in aldolase-catalyzed reactions: the enzymes are specific for the donor substrate but flexible for the acceptor substrate, and the stereochemistries of the aldol addition reactions are controlled by the enzymes, not by the substrates. This review describes some of our recent work on the use of three microbial aldolases for the synthesis of novel sugars and derivatives.

Recombinant Fructose-1,6-Diphosphate (FDP) Aldolase from *E. Coli*

The type I Schiff-base forming FDP aldolase from rabbit muscle is commercially available and is the most often used aldolase in organic synthesis (5,8-11). The type II zinc-containing FDP aldolase from microorganisms, however, has not been used in preparative synthesis. To explore the synthetic utility of the zinc-containing FDP aldolase, we have recently conducted the cloning, overexpression and characterization of the enzyme from *E. Coli* (12). A typical 6-L fermentation of the recombinant *E. Coli* would produce 30 g of crude extract, which contain approximately 120,000 units of FDP aldolase (1 unit of enzyme will catalyze the cleavage of 1 μmol of FDP per minute). The enzyme is a dimer with a molecular weight of ~80,000. The protein sequence has been determined based on the cDNA sequence (13). It is more stable than the rabbit enzyme. In the presence of

0097–6156/91/0466–0023$06.00/0

0.3 mM Zn^{++}, the half life at 25°C is about 60 days compared to 2.5 days for the rabbit enzyme (12). The active-site zinc ion can be replaced with Co^{++}. The Co^{++}-containing enzyme, however, is slightly less active and less stable than the zinc enzyme (14). Figure 1 illustrates the reaction mechanisms of both types of FDP aldolases and the stability and activity of the Co^{++}- and Zn^{++}-containing aldolases. Preliminary studies indicate that the Zn^{++}-containing aldolase has similar substrate specificity to the rabbit muscle enzyme. The two enzymes, however, have little homology (15-16). The molecular architecture of the rabbit enzyme has been determined to 2.7 Å (17). Work is in progress to determine the X-ray crystal structure of the zinc-containing enzyme as an effort for the next stage of research to engineer the substrate specificity.

Azasugars

Piperidines structurally related to monosaccharides are useful as glycosidase inhibitors. We have developed a chemo-enzymatic strategy for the synthesis of this type of molecule based on FDP aldolase (10,12). As shown by the retrosynthetic analysis in Figure 2, many azasugars perhaps can be prepared based on an aldolase reaction followed by a catalytic reductive amination. For the synthesis of deoxynojirimycin and deoxymannojirimycin using FDP aldolase, the reactions were very successful with an overall yield of 50-75% for each compound. The reductive aminations were found to be regioselective, with only one diastereomer formed. Hydrogenation of the imine intermediate always gave the product with a trans C_4-C_5 relation. A similar strategy may be applied to the two other aldolases which use dihydroxyacetone phosphate as the donor substrate. Figure 3 outlines the preparation of enantiomerically pure aldehydes via a lipase reaction for use in the FDP aldolase reactions. We have also extended the combined aldol reaction and reductive amination strategy to the synthesis of azasugars corresponding to N-acetylglucosamine and N-acetylmannosamine (Figure 4). The aldehyde substrates were prepared from 3-azido-2-hydroxypropanal diethyl acetal as outlined in Figure 4. Nucleophilic opening of the aziridine with sodium azide in the presence of zinc triflate was found to be quite satisfactory (~40% yield). These azasugars are being evaluated as inhibitors of glycosidases involved in di- and oligosaccharide degradation and glycoprotein processing. Azasugars hold potential value for the treatment of symptoms or diseases associated with these biochemical processs.

Conversion of Ketoses to Aldoses

Some of the ketoses prepared in the FDP aldolase reactions can be converted to the corresponding aldolases catalyzed by glucose isomerase (8) (from Miles), the enzyme used for the manufacture of high fructose corn syrup. Figure 5 summarizes some of the isomerization reactions. The equilibrium mixture can be separated by Dowex-50-Ba^{++} column chromatography using water as the mobile phase. This isomerization step provides a new entry to several interesting aldolases.

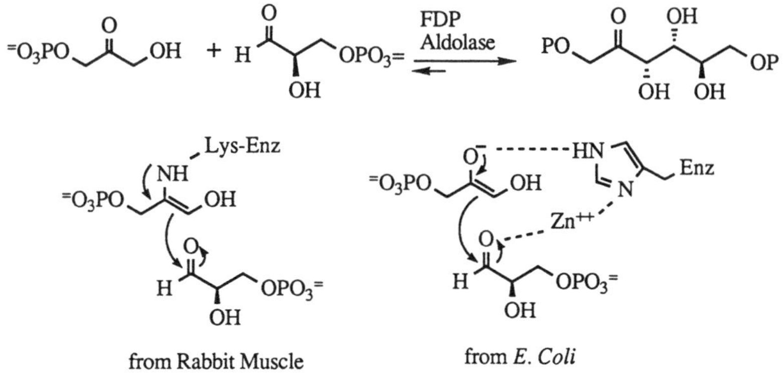

$$E\,Zn_2 \rightleftharpoons E\,Zn + Zn(II) \quad K_1 = 10^{-8.9}\,M \qquad E\,Co_2 \rightleftharpoons E\,Co + Co(II) \quad K_1 = 10^{-7.6}M$$

$$E\,Zn \rightleftharpoons E + Zn(II) \quad K_2 = 10^{-11.8}\,M \qquad E\,Co \rightleftharpoons E + Co(II) \quad K_2 = 10^{-10.9}M$$

$$t_{1/2}^{60^\circ C} = 1.5\text{ h} \quad V_{max} = 25\text{U/mg} \qquad t_{1/2}^{60^\circ C} = 0.5\text{ h} \quad V_{max} = 20\text{ U/mg}$$

Figure 1. (Top) FDP Aldolase from rabbit muscle and *E. Coli.* (Bottom) Activity, stability and dissociation constants of Zn^{++}- and Co^{++}-FDP aldolase from *E. Coli.* at 20°C, pH 8.

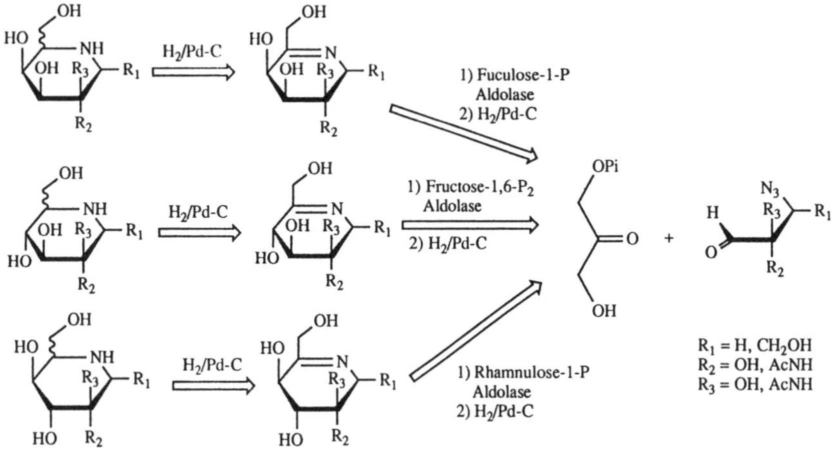

Figure 2. Retrosynthetic analysis for the synthesis of azasugars.

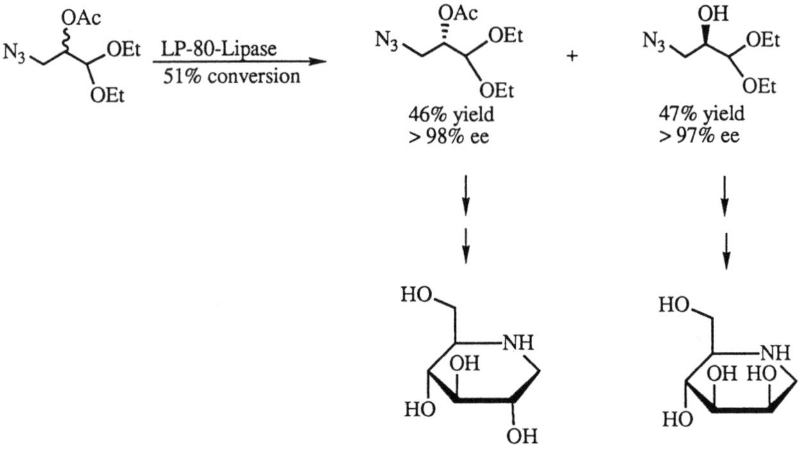

Figure 3.   Syntheses of deoxynojirimycin and deoxymannojirimycin.

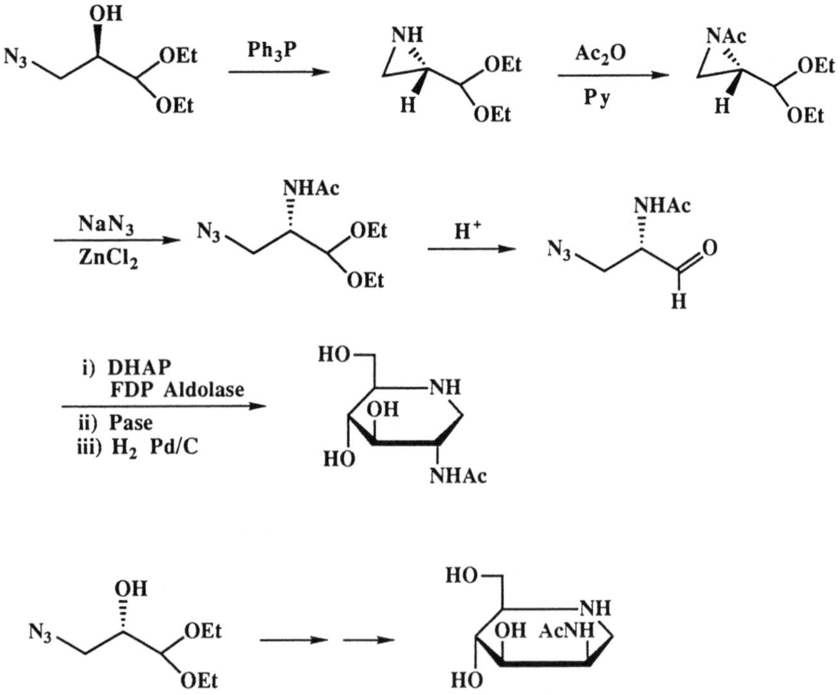

Figure 4.   Syntheses of the azasugars corresponding to N-acetylgluco-samine and N-acetylmannosamine.

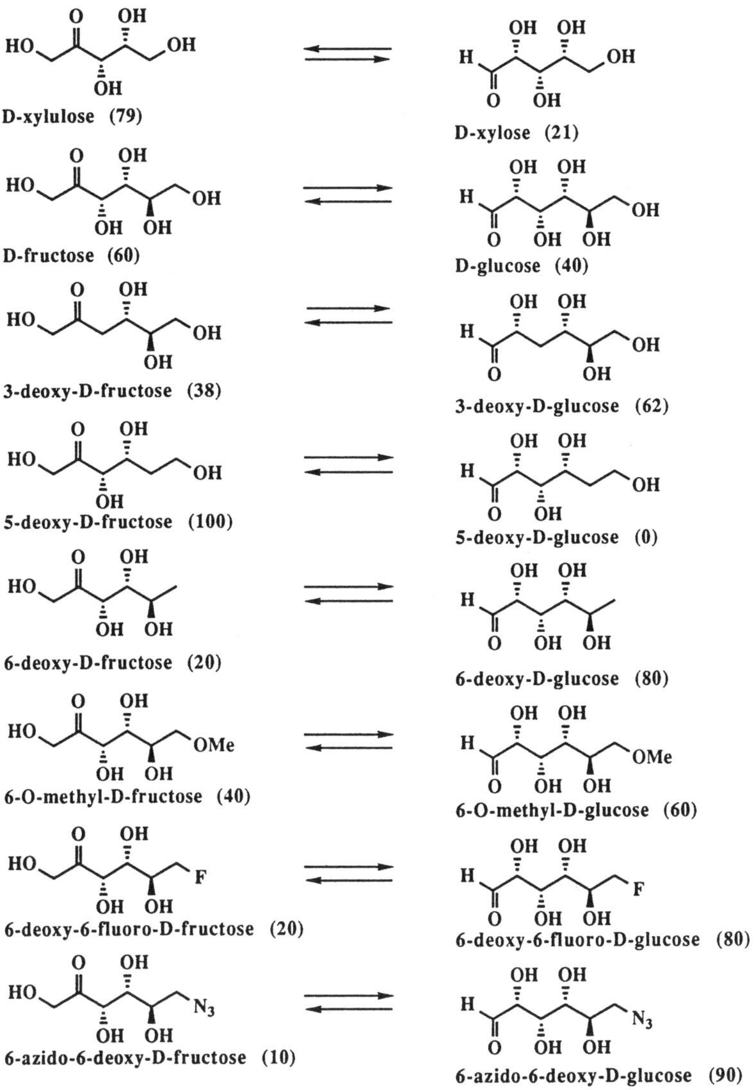

D-xylulose (79) ⇌ D-xylose (21)

D-fructose (60) ⇌ D-glucose (40)

3-deoxy-D-fructose (38) ⇌ 3-deoxy-D-glucose (62)

5-deoxy-D-fructose (100) ⇌ 5-deoxy-D-glucose (0)

6-deoxy-D-fructose (20) ⇌ 6-deoxy-D-glucose (80)

6-O-methyl-D-fructose (40) ⇌ 6-O-methyl-D-glucose (60)

6-deoxy-6-fluoro-D-fructose (20) ⇌ 6-deoxy-6-fluoro-D-glucose (80)

6-azido-6-deoxy-D-fructose (10) ⇌ 6-azido-6-deoxy-D-glucose (90)

The following are not substrates: D-allose, D-galactose, 4-deoxy-glucose, 4-deoxy-fluoroglucose, 5-thio-D-glucose, L-rhamnose, L-fucose, L-sorbose.

Figure 5. Isomerization of the products from FDP aldolase reactions to aldolases catalyzed by glucose isomerase. The numbers in parenthesis indicate the equilibrium ratio.

Dihydroxyacetone   phosphate

The FDP aldolase and two other aldolases require dihydroxyacetone phosphate as a substrate.   Current methods for the preparation of dihydroxyacetone phosphate are from dihydroxyacetone by chemical or enzymatic synthesis, and from FDP by *in situ* enzymatic generation (*6*). Dihydroxyacetone phosphate can also be replaced with dihydroxyacetone and catalytic amounts of inorganic arsenate (*18*).
   We have recently developed an improved procedure for the chemical synthesis which is outlined in Figure 6.   The major improvement is the use of a trivalent phosphorylating reagent, dibenzyl-N, *N*-d i e t h y l phosphoramidite, in the presence of triazole or tetrazole.   The phosphorylating reagent was easily prepared from PCl$_3$, benzyl alcohol and diethylamine.   The overall yield of dihydroxyacetone phosphate from dihydroxyacetone is approximately 55%.   The process does not require chromatography and the intermediates are crystalline and easy to handle. This phosphorylating procedure is being extended to the synthesis of other sugar phosphates.

Recombinant   2-Deoxyribose-5-phosphate   Aldolase   (DERA)   from
E. Coli

This enzyme catalyzes the aldol condensation between acetaldehyde and D-glyceraldehyde-3-phosphate to 2-deoxyribose-5-phosphate.   It is the only aldolase that accepts two aldehydes as substrates.   It has been cloned and overexpressed in *E. Coli* (Figure 7) (*19*).   A 6-L fermentation produced approximately 124,000 units of the enzyme.   The purified enzyme showed optimal activity at pH 7.5 with V_{max} = 55 U/mg.   Examination of the substrate specificity of the enzyme indicates that acetone, fluoroacetone, and propionaldehyde are also accepted by the enzyme as donor substrates. The rates for the condensations, however, are about 1% that of the natural reaction.   Chloroacetaldehyde and glycoaldehyde are not donor substrates. It is interesting that the enzyme does not catalyze the self-condensation of acetaldehyde but does for propionaldehyde.   With fluoroacetone as donor, the reaction occurs regioselectively at the nonfluorinated carbon.   Figure 8 illustrates representative syntheses (*19-20*).   In the synthesis of 2-deoxyribose-5-phosphate, D-glyceraldehyde-3-phosphate was generated *i n situ* from dihydroxyacetone phosphate catalyzed by triosephosphate isomerase.   Alternatively, it can be prepared from (R)-glycidaldehyde diethyl acetal via a lipase resolution.   Work is in progress to extend the use of this enzyme to the synthesis of other β-hydroxy compounds and to further investigate the substrate specificity of the enzyme.
   During the preparation of a recombinant plasmid for DERA, we also subcloned the genes coding for thymidine phosphorylase and phosphopentomutase from the plasmid pVH17 and ligated in the pKK expression vector for the expression of these two enzymes (*21*) (Figure 7).

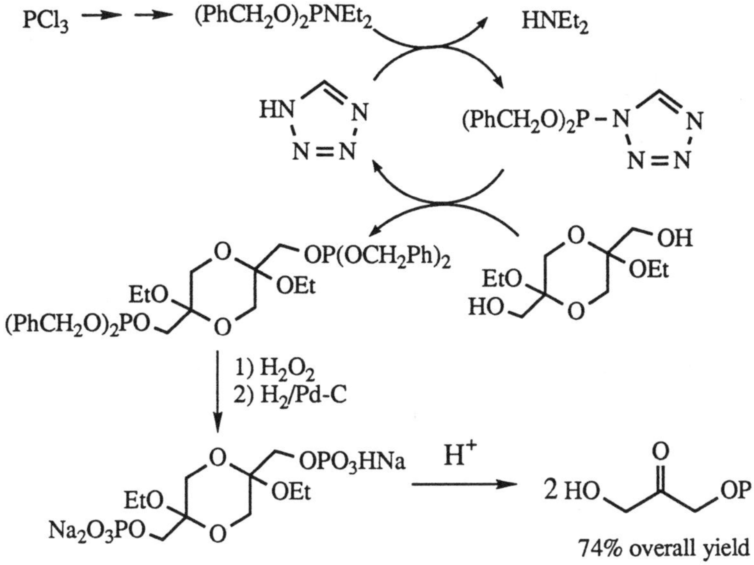

Figure 6.   Synthesis of dihydroxyacetone phosphate.

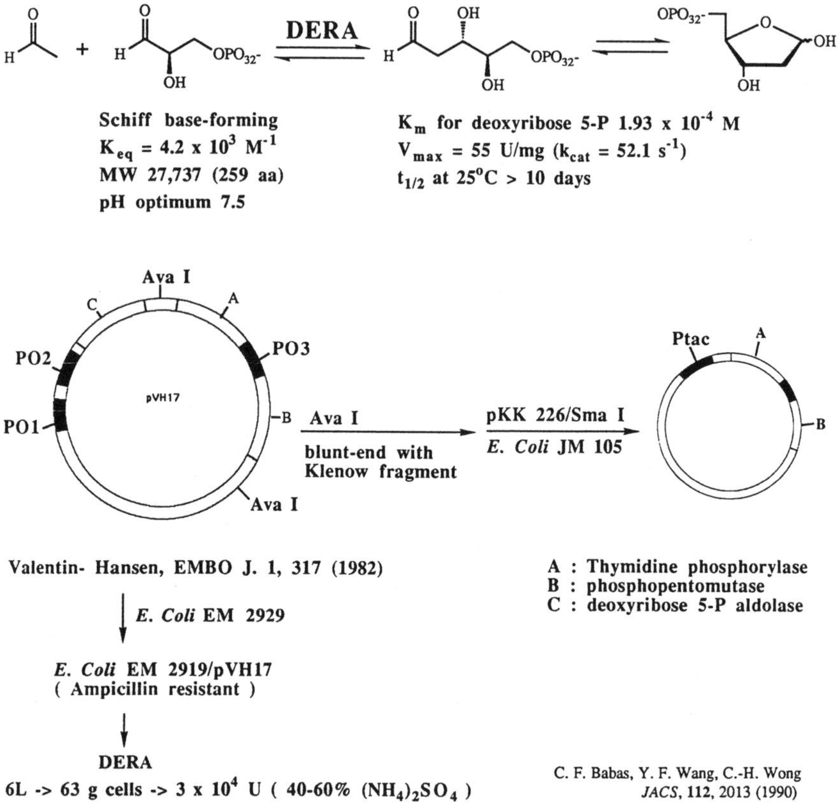
DERA
Schiff base-forming
K_eq = 4.2 x 10^3 M^-1
MW 27,737 (259 aa)
pH optimum 7.5
K_m for deoxyribose 5-P 1.93 x 10^-4 M
V_max = 55 U/mg (k_cat = 52.1 s^-1)
t_1/2 at 25°C > 10 days
Ava I
C
A
PO2
PO3
pVH17
PO1
B
Ava I
Ava I
blunt-end with
Klenow fragment
pKK 226/Sma I
E. Coli JM 105
Ptac
A
B
Valentin- Hansen, EMBO J. 1, 317 (1982)
E. Coli EM 2929
E. Coli EM 2919/pVH17
(Ampicillin resistant)
DERA
6L -> 63 g cells -> 3 x 10^4 U (40-60% (NH4)2SO4)
A : Thymidine phosphorylase
B : phosphopentomutase
C : deoxyribose 5-P aldolase
C. F. Babas, Y. F. Wang, C.-H. Wong
JACS, 112, 2013 (1990)

Figure 7. Plasmid construction for the preparation of recombinant deoxyribose-5-phosphate aldolase (DERA), phosphopentomutase and nucleoside phosphorylase.

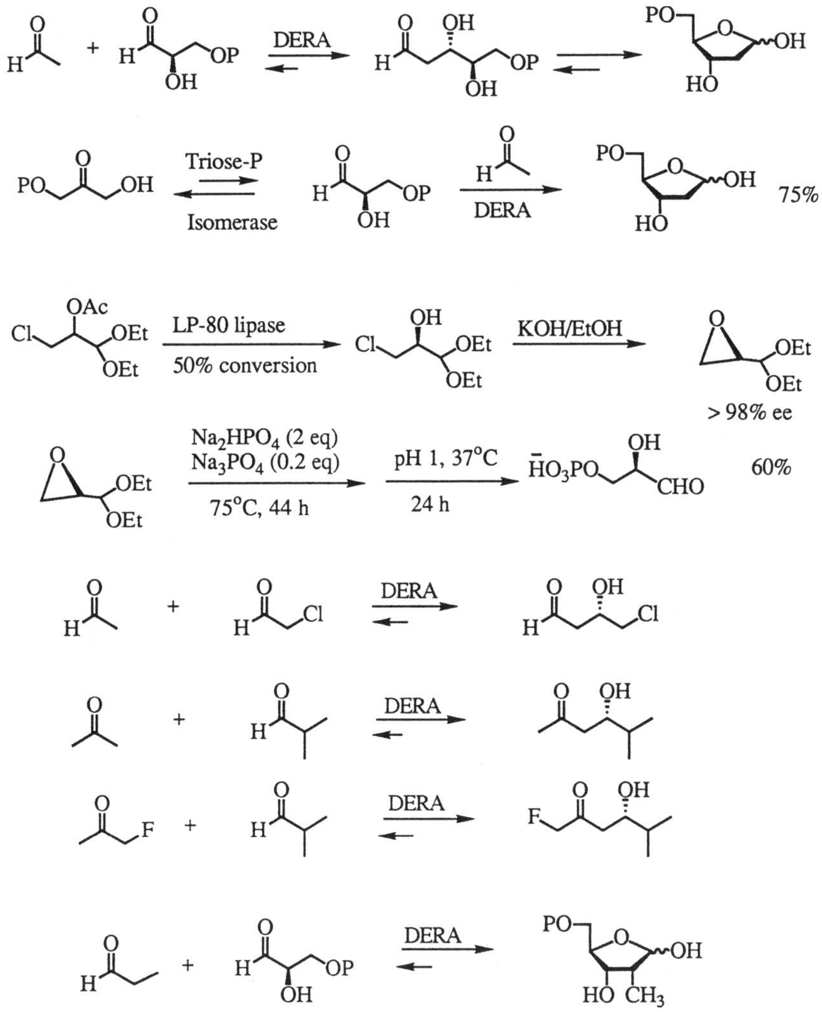

Figure 8.   Representative syntheses using DERA.

Initial studies indicate that phosphopentomutase also catalyzes the isomerization of ribose-5-phosphate and arabinose-5-phosphate, in addition to 2-deoxyribose-5-phosphate, to the corresponding α-1-phosphates; 2,3-Dideoxyribose-5-phosphate, however, was not a substrate. Coupling of the α-1-phosphate *in situ* with a purine or a pyrimidine base catalyzed by purine nucleoside phosphorylase or pyrimidine nucleoside phosphorylase gives a nucleoside *(21)*. Given the fact that α-pentose-1-phosphate is unstable and difficult to prepare and that nucleoside phosphorylase also accepts a broad range of substrates, this coupling enzymatic system may provide a new opportunity for the synthesis of novel nucleosides with antiviral activities. 2-Deoxyadenosine and Virazole, for example, have been prepared based on the coupling reactions (Figure 9).

N-Acetylneuraminic Acid Aldolase (Sialic Acid Aldolase)

This commercially available enzyme (from Toyobo) has been used for the synthesis of a number of sugars structurally related to sialic acid *(22,23)*. Figure 10 summarizes studies on the substrate specificity and synthetic application of the enzyme *(22-24)*. Our recent effort has been focused on the development of new procedures for the preparation of acceptor substrates for the aldolase reaction. Selective deprotection of peracylated glycosides at the primary position with lipases or subtilisin allows for the preparation of hexoses with modification at the position C6 *(25)*. Regioselective acylation of hexoses at the primary position with either the wild-type subtilisin or a genetically-engineered thermostable subtilisin *(26)* followed by the aldolase reaction provided a new entry to 9-0-acylsialic acids *(23)*. 9-0-Acetylsialic acid, for example, was prepared in a two-step reaction with an overall yield of 82% *(23)*. Similarly, 9-0-L-lactylsialic acid has recently been prepared *(27)*. Sialic acid has also been converted to α- and β-2-deoxysialic acid with a reasonably high yield (Figure 11).

Conclusion

The enzyme-catalyzed aldol condensation is obviously a useful technology for the synthesis of natural and many unnatural sugars. With the cloning technology available for the preparation of enzymes, many other aldolases will continue to be explored for synthesis. It is conceivable that aldolases and their products will constitute a new group of synthetic catalysts and chiral intermediates which will be used substantially for the synthesis of polyhydroxy compounds. The increasing availability of aldolases and glycosyltransferases *(29)* through cloning techniques should have a significant impact on carbohydrate synthesis in the near future.

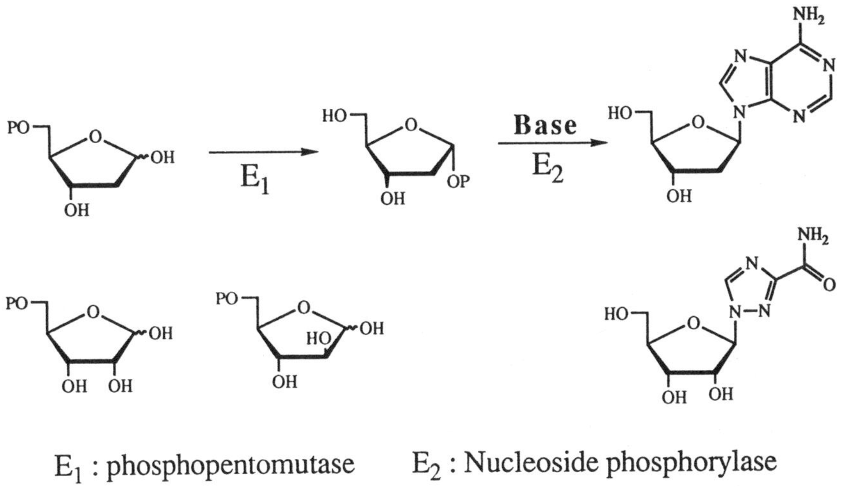

E_1 : phosphopentomutase E_2 : Nucleoside phosphorylase

Figure 9. Coupling of phosphopentomutase with nucleoside phosphorylase.

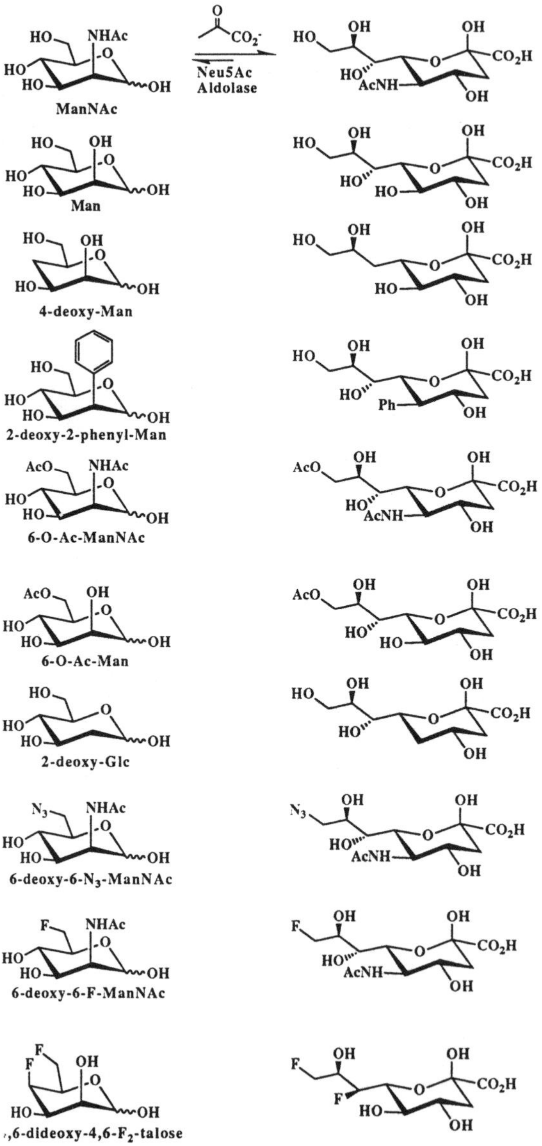

Figure 10. Sialic acid aldolase-catalyzed reactions.

	Yield	Rel V	K_M (Acceptor)
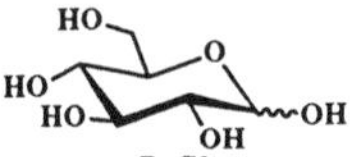 D-Glc	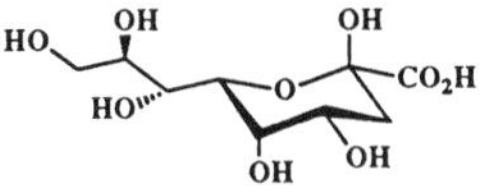 —	0.07	2.3 M
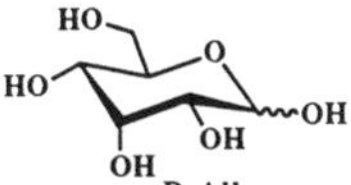 D-All	—	—	0.004	—
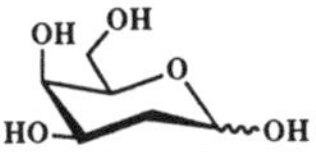 2-deoxy-Gal	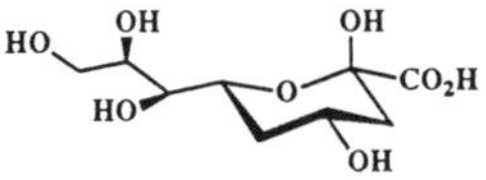 —	0.2	1.3 M
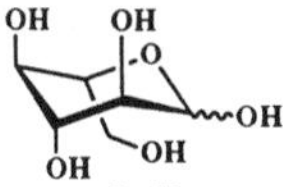 L-Glc	—	—	0	—
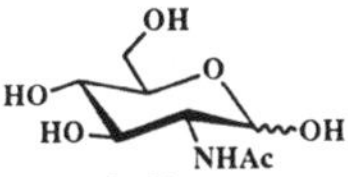 GlcNAc	—	—	0	—
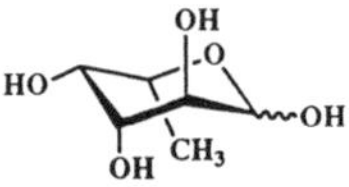 L-Fucose	—	—	0.04	—
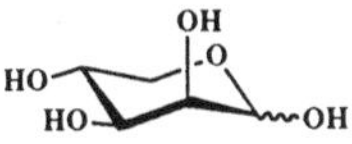 D-Lyxose	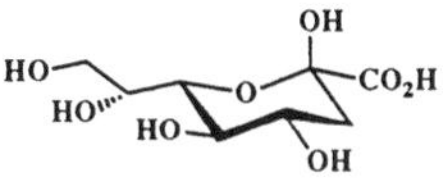 36%	0.13	1.7 M
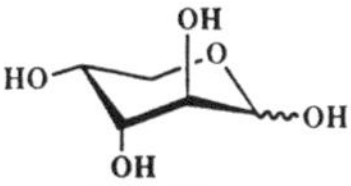 D-Arabinose	—	—	0.03	—
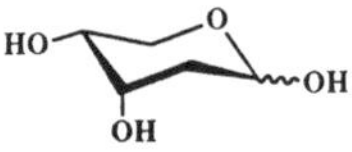 2-deoxy-ribose	—	—	0.02	—
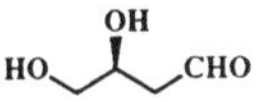	—	—	0	—
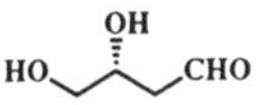	—	—	0	—

Figure 10 (*continued*).

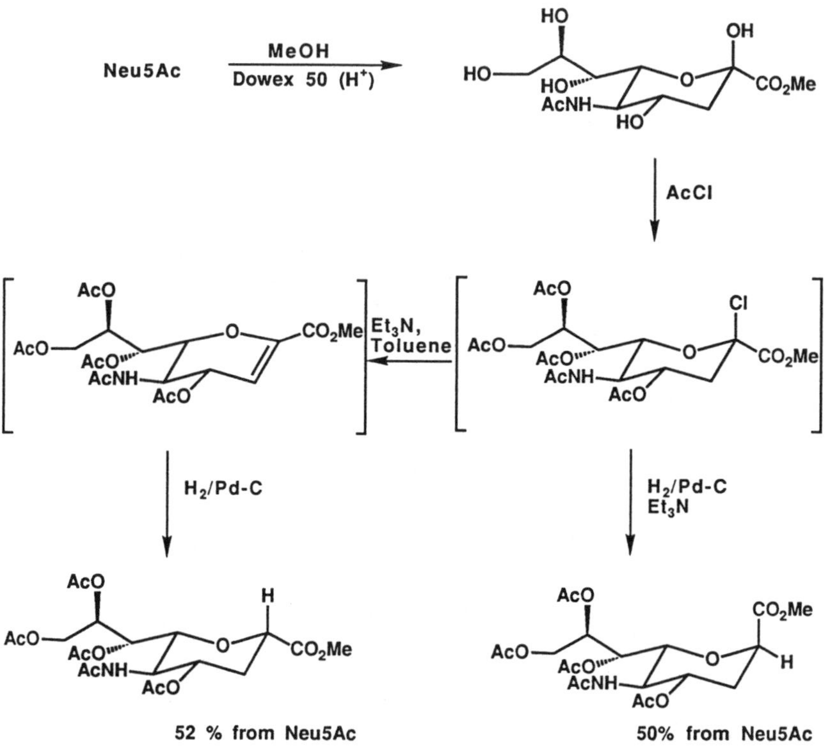

Figure 11.   Syntheses of α- and β-2-deoxysialic acid.

Acknowledgments

This work was carried out by a number of members whose names are listed in the references.   The work was funded by the NIH GM44154-01.

Literature   Cited

1. Bednarski, M.; Maring, C.; Danishefsky, S. *Tetrahedron Lett.* **1983**, *24*, 3451.
2. Ando, A.; Shioiri, T. *J. Chem. Soc. Chem. Comm.* **1987**, 1620.
3. Nakagawa, M.; Nakao, H.; Watanabe, K. *Chem. Lett.* **1985**, 391.
4. Ito, Y.; Sawamura, M.; Hayashi, T. *J. Am. Chem. Soc.* **1986**, *108*, 6405.
5. Bednarski, M. D.; Simon, E. S.; Bischofberger, N.; Fessner, W. -D.; Kim, M. -J.; Lees, W.; Saito, T.; Waldmann, H.; Whitesides, G. M. *J. Am. Chem. Soc.* **1989**, *111*, 627.

6. Toone, E. J.; Simon, E. S.; Bednarski, M. D.; Whitesides, G. M. *Tetrahedron Lett.* **1989**, *45*, 6355.
7. Wong, C. -H.; Drueckhammer, D. G.; Durrwachter, J. R.; Lacher, B.; Chauvet, C. J.; Wang, Y. -F.; Sweers, H. M.; Smith, G. L.; Yang, J. -S.; Hennen, W. J. In *Trends in Synthetic Carbohydrate Chemistry*; Horton, D.; Hawkins, L. D.; McGarvey, G. J., Eds.; ACS Symposium Series No. 386; American Chemical Society: Washington, D.C. 1989, pp 317-335.
8. Durrwachter, J. R.; Drueckhammer, D. G.; Nozaki, K.; Sweers, H. M.; Wong, C. -H. *J. Am. Chem. Soc.* **1986**, *108*, 7812.
9. Durrwachter, J. R.; Wong, C. -H. *J. Org. Chem.* **1988**, *53*, 4175.
10. Pederson, R. L.; Kim, M. -J.; Wong, C. -H. *Tetrahedron Lett.* **1988**, *29*, 4645.
11. Ziegler, T.; Straub, A.; Effenberger, F. *Angew. Chem. Int. Ed. Engl.* **1988**, *27*, 716.
12. von der Osten, C. H.; Sinskey, A. J.; Barbas, C. F., III; Pederson, R. L.; Wang, Y. -F.; Wong, C. -H. *J. Am. Chem. Soc.* **1989**, *111*, 3924.
13. Alefounder, P. R.; Baldwin, S. A.; Perham, R. N.; Short, N. J. *Biochem. J.* **1989**, *257*, 529.
14. Chen, Y. L. Metalloenzyme Chemistry, Master Thesis, Texas A&M University, 1989.
15. von der Osten, C. H.; Barbas, C. F., III; Wong, C. -H.; Sinskey, A. J. *Molecular Microbiology* **1989**, *3*, 1625.
16. Malek, A. A.; Yu, M.; Honegger, A.; Rose, K.; Brenner-Holzach, O. *Arch. Biochem. Biophys.* **1988**, *266*, 10.
17. Sygusch, J.; Beaudry, D.; Allaire, M. *Proc. Natl. Acad. Sci.* **1987**, *84*, 7846.
18. Drueckhammer, D. G.; Durrwachter, J. R.; Pederson, R. L.; Crans, D. C.; Wong, C. -H. *J. Org. Chem.* **1989**, *54*, 70.
19. Barbas, C. F., III; Wang, Y. -F.; Wong, C. -H. *J. Am. Chem. Soc.* **1990**, *112*, 2013.
20. Chen, L.; Wong, C. -H., unpublished data.
21. Barbas, C. F., III. Overproduction and Utilization of Enzymes in Synthtic Organic Chemistry, Ph.D. Thesis, Texas A&M University, 1989.
22. Auge, C.; David, S.; Gautheron, C.; Malleron, A.; Cavaye, B. *New J. Chem.* **1988**, *12*, 733.
23. Kim, M. -J.; Hennen, W. J.; Sweers, H. M.; Wong, C. -H. *J. Am. Chem. Soc.* **1988**, *110*, 6481.
24. Auge, C.; Gautheron, C.; David, S. *Tetrahedron* **1990**, *46*, 201.
25. Hennen, W. J.; Sweers, H. M.; Wang, Y. -F.; Wong, C. -H. *J. Org. Chem.* **1988**, *53*, 4939.
26. Wong, C. -H.; Chen, S. -T.; Hennen, W. J.; Bibbs, J. A.; Wang, Y. -F.; Liu, J. L. -C.; Pantoliano, M. W.; Whitlow, M.; Bryan, P. N. *J. Am. Chem. Soc.* **1990**, *112*, 945.
27. Liu, J. L. -C.; Wong, C. -H., unpublished data.
28. Pederon, R. L.; Wong, C. -H., submitted for publication.
29. Paulson, J. C.; Colley, K. C. *J. Biol. Chem.* **1989**, *264*, 17615.

RECEIVED February 7, 1991

Chapter 3

Use of Glycosyltransferases in Synthesis of Unnatural Oligosaccharide Analogs

Ole Hindsgaul[1], Kanwal J. Kaur[1], Uday B. Gokhale[1], Geeta Srivastava[1], Gordon Alton[1], and Monica M. Palcic[2]

Departments of [1]Chemistry and [2]Food Science, University of Alberta, Edmonton, Alberta T6G 2G2, Canada

The general ability of glycosyltransferases to utilize analogs of the natural sugar-nucleotides as donor substrates is demonstrated. Specifically, the human Lewis $\alpha(1\rightarrow4)$ fucosyltransferase was found to transfer both 3-deoxy-L-fucose and D-arabinose from their GDP-derivatives to $\beta Gal(1\rightarrow3)\beta GlcNAc$-$O(CH_2)_8COOMe$. Both N-acetylglucosaminyltransferases I and II were able to utilize UDP-3,4 and 6-deoxy-GlcNAc as donors to produce deoxygenated oligosaccharide analogs on a mg scale. UDP-3-deoxy-Gal was also a donor substrate for $\beta(1\rightarrow4)$ galactosyltransferase. The preparation of oligosaccharide analogs by a combined chemical-enzymatic approach is proposed as an alternative to their more laborious total chemical synthesis.

The complex carbohydrate chains of mammalian glycoproteins and glycolipids became important targets for organic synthesis after their structural characterization as the antigenic determinants of the human ABO blood group system in the 1960's (1). Since that time, oligosaccharide structures have been shown to function as recognition molecules in a diversity of biological roles including those of receptors for the binding of bacteria (2-4) and viruses (5), as developmental markers or as tumor associated antigens (6-9). The oligosaccharides mediating these important biological phenomena are usually between two and five or six sugar residues in size. In general, these compounds are difficult to purify from natural sources in quantities sufficient for systematic biochemical studies such as the determination of association constants or the preparation of neoglycoproteins and affinity matrices.

Many elegant methods (10-12) have been developed for the chemical synthesis of biologically significant oligosaccharides in response to the above noted needs for structurally well-defined molecules. As a result of these developments in chemical synthesis, penta- and hexasaccharides have been increasingly referred to as "routine" synthetic targets. While this may occasionally be true, judging from the experience in our laboratory, the "routine" preparation of chemically-synthesized oligosaccharides requires an average of 7 steps per monosaccharide residue taking about one week of work per step. This means that an average pentasaccharide will take about 35 weeks to prepare in the hands of an experienced individual. While oligosaccharides of this size can frequently be prepared much more rapidly, it is equally true that some can take much longer to synthesize and can sometimes not even be completed. The above "average" includes all of these cases. The compound will therefore cost approximately

0097–6156/91/0466–0038$06.00/0

$30,000 in grant support to prepare in an academic laboratory (in Canada) and closer to $75,000 in an industrial laboratory. The chemical synthesis of oligosaccharides should therefore be considered "routine" only within these constraints of personnel and money. Unfortunately, such resources are available to few research groups with the result that "longshot-leads" are rarely followed up on. For example, the chemical synthesis of a new oligosaccharide predicted to possess a novel biological activity is an extremely risky venture because the cost leaves no allowance for failure either in the primary synthesis or subsequent biological testing. To synthesize and evaluate a panel of analogs based on tentative leads is therefore usually prohibitive in terms of cost and labour.

Glycosyltransferases transfer sugars from their activated forms of sugar-nucleotides to oligosaccharide acceptors which can be as small as mono- to trisaccharides (Figure 1). The use of glycosyltransferases in a combined chemical-enzymatic approach to the preparation of oligosaccharides has the potential for enormous savings in both time and money. The advantage of using these enzymes in the formation of glycosidic linkages is that they are both stereospecific and regiospecific, obviating the need for laborious protection-deprotection strategies (13). The seven major sugar-nucleotides needed for the synthesis of mammalian glycoconjugates, i.e. UDP-Glc, UDP-Gal, UDP-GlcNAc, UDP-GalNAc, GDP-Man, GDP-Fuc and CMP-NeuAc, are now commercially available although they are expensive. They can also be prepared on large scales using enzymatic procedures (13). The two most serious impediments to the broad use of glycosyltransferases in preparative oligosaccharide synthesis therefore appear to be:
a) Availability of the approximately 100-150 enzymes required for the biosynthesis of known oligosaccharide sequences and
b) The anticipation that inherently poor reactivity with modified substrates will severely limit the use of the enzymes in the preparation of chemically modified oligosaccharide analogs.

With regard to the question of enzyme availability, many examples have been reported where glycosyltransferases, some only partially purified, have been used in the production of oligosaccharides on the milligram and even multigram scales (13). In addition, phenomenal progress has been made in the last few years in the molecular biology of glycosyltransferases (14) with the result that about a dozen of these enzymes have now been cloned. It can therefore be anticipated that a large number of the required glycosyltransferases will shortly become available, either commercially or through scientific collaboration. We have therefore turned our attention to point **b** above, namely defining the scope and utility of glycosyltransferases in the preparation of unnatural oligosaccharide analogs.

We report here on our recent progress in using glycosyltransferases mainly in the preparation of deoxygenated oligosaccharides. Our particular emphasis on deoxygenated compounds, as opposed to halogenated or other analogs, stems from a wide body of literature demonstrating the utility of such compounds in probing the the molecular specificity of carbohydrate-protein recognition (15). The realization that deoxygenated analogs of glycosyltransferase acceptors can also serve as potent inhibitors for the enzymes (16) has further increased the potential use of such analogs. Before describing the results from our laboratory, we wish to explicitly point out previous published work in this area which further demonstrates the ability of glycosyltransferases to transfer modified sugar residues. In particular, $\beta(1\rightarrow4)$galactosyltransferase has been shown to transfer D-glucose, 4-deoxy-D-glucose and L-arabinose from their UDP-derivatives (17) and a mannosyltransferase similarly transfers 3,4 and 6-deoxy-D-mannose from their GDP-nucleotides (18). An extensive literature exists on the preparation of CMP-sialic acid analogs and their utilization by sialyltransferases (19). Russian workers (21,22) have also shown that bacterial glycosyltransferases can transfer deoxygenated analogs of the naturally-occuring sugars to suitable acceptors.

The first enzyme we evaluated for its ability to transfer unnatural sugars was the human Lewis $\alpha(1\rightarrow4)$fucosyltransferase (FucT). This particular enzyme was chosen because of its potential utility in the synthesis of the tumor associated sialyl-Lewis-a antigen (22) and analogs thereof. In addition, the enzyme could be quite readily obtained in a soluble form with high specific activity from human milk. The $\alpha(1\rightarrow4)$FucT catalyzes reaction I below which results in the production of the Lewis-a trisaccharide determinant.

$$\text{(I) } \beta Gal(1\rightarrow3)\beta GlcNAc\text{-OR} + \text{GDP-Fuc} \xrightarrow{\text{FucT}} \overset{\alpha Fuc(1,4)\downarrow}{\beta Gal(1\rightarrow3)\beta GlcNAc\text{-OR}} + \text{GDP}$$

Simple chemical syntheses of GDP-fucose (1) and its analogs 2 and 3 were devised which involved the direct reaction of α-pyranosyl bromides with tetrabutylammonium dibenzylphosphate (Figure 2) (23). These displacement reactions were carried out with protected derivatives of the parent L-fucose, 3-deoxy-L-fucose and D-arabinose (the 6-demethyl analog of L-fucose) because of the relative ease of preparation of the required glycosyl bromides. The β-phosphates thus obtained were coupled with GMP-morpholidate to yield the GDP derivatives 1 - 3.

Sugar nucleotides 1- 3 were found to act as donors for the $\alpha(1\rightarrow4)$FucT. Their relative rates of reaction were estimated using a novel spectrophotometric coupled enzyme assay (23) which correlated the rate of production of GDP (equation I) with the rate of glycosyl transfer. The results, presented in Figure 3, show that 3-deoxy-L-fucose and D-arabinose were transferred at 1.7 and 6.0%, respectively, of the rate for L-fucose. The structures of the products were identified by an ELISA assay and scaled-up reactions (≈2 mg) allowed the product trisaccharides to be characterized by NMR (23). The natural Lewis-a trisaccharide determinant as well as two of its analogs could thus be prepared enzymatically.

We have also examined the utility of N-acetylglucosaminyltransferases I and II in the synthesis of analogs of asparagine-linked (Asn-linked) oligosaccharides. These enzymes transfer βGlcNAc residues from UDP-GlcNAc to the core trimannoside of Asn-linked oligosaccharides as shown in equation II. In the natural biosynthetic pathway, GlcNAcT-I converts the minimum structure 4 to hexasaccharide 5 which is further homologated to heptasaccharide 6 by GlcNAcT-II (24). We have found that the chitobiose unit of the natural acceptors can be substituted by aliphatic aglycones and the resulting oligosaccharides 7 and 8 still retain excellent acceptor properties for these two enzymes (25,26).

The 3,4 and 6-deoxy-analogs of UDP-GlcNAc were chemically synthesized (26) from the corresponding peracetylated reducing sugars, using the method of Inage et al. (27) as summarized in Figure 4. The α-phosphates obtained in this manner were coupled, as before, with UMP-morpholidate to yield the UDP-deoxy-GlcNAc analogs 10 - 12. All three analogs were compared as donor substrates for GlcNAcT-I, partially purified from human milk, using trisaccharide 7 as the acceptor. The relative rates of glycosyl-transfer have not yet been determined for these analogs. Instead,

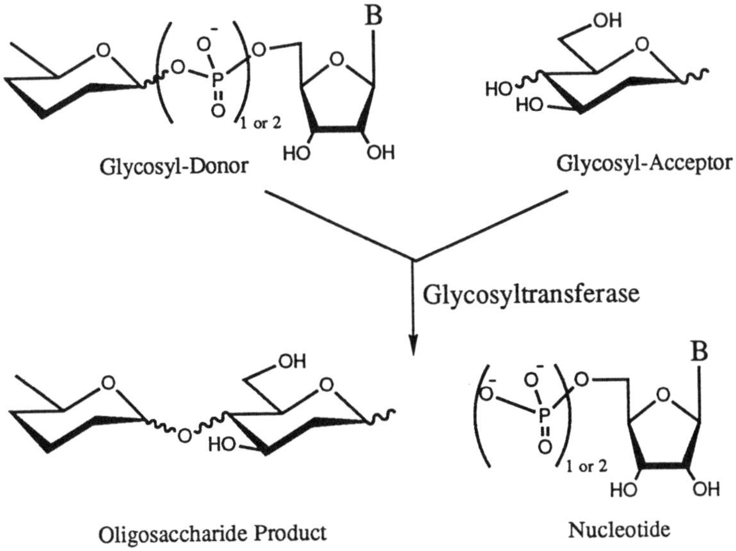

Figure 1. A Generalized Scheme for Glycosyltransferase Reactions.

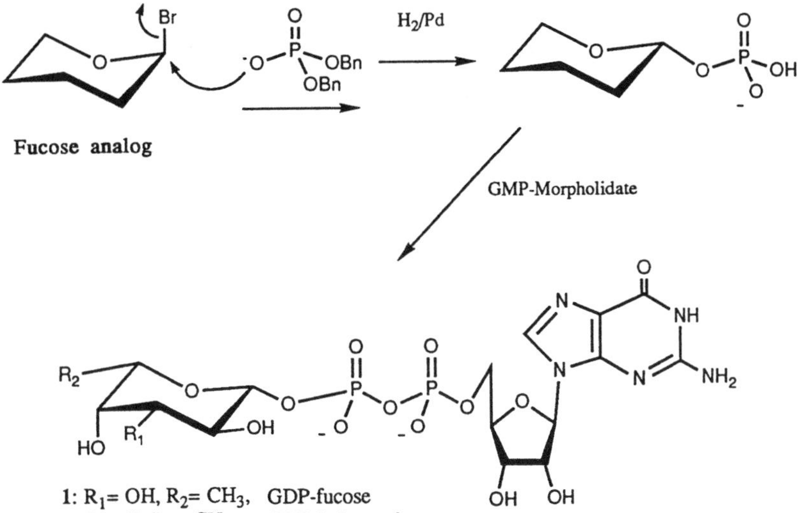

1: R_1= OH, R_2= CH_3,   GDP-fucose
2: R_1= H, R_2= CH_3,   GDP-3-deoxy-fucose
3: R_1= OH, R_2=H,   GDP-arabinose

Figure 2. Synthesis of GDP-Fucose and Analogs.

βGal(1→3)βGlcNAc-O(CH$_2$)$_8$COOMe

GDP-Fucose
Analogs   FucT (human milk)
1 - 3

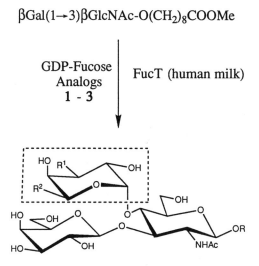

GDP-Fuc Analog	Product	Rel. Rate	% Yield
1 →	Le^a (R^1=OH,R^2=Me)	100	100
2 →	3"-deoxy-Le^a (R^1=H,R^2=Me)	1.7	25
3 →	6"-demethyl-Le^a (R^1=OH, R^2=H)	6.0	80

Figure 3. Transfer of Fucose Analogs by the Lewis α(1→4)
Fucosyltransferase.

αMan(1→6)
⟩βMan-R
αMan(1→3)

4: R = βGlcNAc(1→4)βGlcNAc-Asn
7: R = O(CH₂)₈COOMe

UDP-GlcNAc | **GlcNAcT-I**

αMan(1→6)
(II) βGlcNAc(1→2)αMan(1→3) ⟩βMan-R

5: R = βGlcNAc(1→4)βGlcNAc-Asn
8: R = O(CH₂)₈COOMe

UDP-GlcNAc | **GlcNAcT-II**

βGlcNAc(1→2)αMan(1→6)
⟩βMan-R
βGlcNAc(1→2)αMan(1→3)

6: R = βGlcNAc(1→4)βGlcNAc-Asn
9: R = O(CH₂)₈COOMe

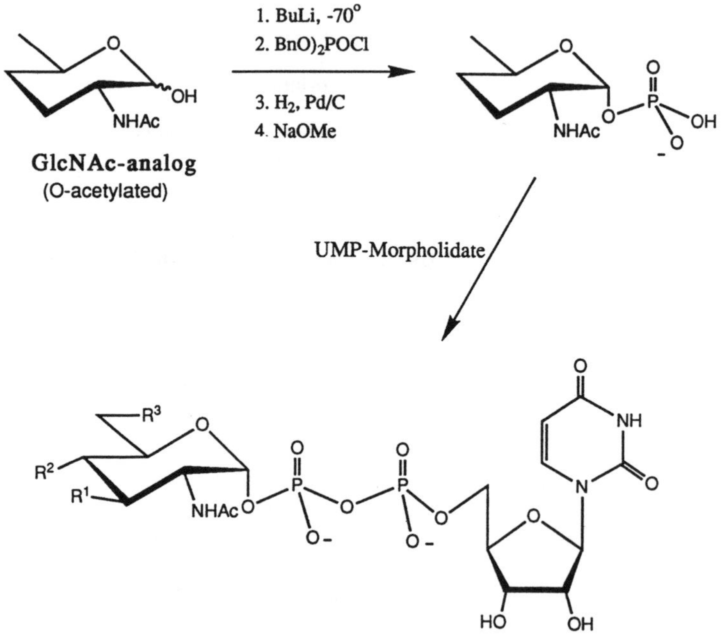

R^1=OH,R^2=OH,R^3=OH,   UDP-GlcNAc
10: R^1=H,R^2=OH,R^3=OH,   UDP-3-deoxy-GlcNAc
11: R^1=OH,R^2=H,R^3=OH,   UDP-4-deoxy-GlcNAc
12: R^1=OH,R^2=OH,R^3=H,   UDP-6-deoxy-GlcNAc

Figure 4. Synthesis of UDP-GlcNAc Analogs.

enzyme incubations were carried out for 48 h at 37°C and the ratio of tetrasaccharide product to unreacted trisaccharide **7** was determined directly from the ^{1}H-NMR spectrum of the unresolved mixture. The deoxy-tetrasaccharides produced in these incubations were separated from unreacted **7** by gel-permeation chromatography on Bio-Gel P-2. The results of these experiments are summarized in Figure 5. All three UDP-GlcNAc analogs were found to be active as donors although the 4-deoxy-GlcNAc derivative **11** was a very poor substrate. Nevertheless, the deoxy-tetrasaccharide products could be isolated in quantities sufficient to permit their evaluation as substrates for GlcNAcT-II. The results of these evaluations have been reported elsewhere (26).

The UDP-deoxy-GlcNAc derivatives **10 - 12** were next evaluated as potential donors for GlcNAcT-II, also partially purified from human milk, using tetrasaccharide **8** as the acceptor. The results are summarized in Figure 6 where it is seen that all three analogs were substrates. The enzyme preparation used was sufficiently active to effect complete transfer of GlcNAc as well as its 3 and 6-deoxy analogs. As in the case of GlcNAcT-I, the rate of transfer of the 4-deoxy derivative **11** was substantially slower than that of the other compounds. The pentasaccharide products could be isolated in 0.5-2.0 mg quantities after chromatography.

It was earlier noted that $\beta(1\rightarrow4)$galactosyltransferase can utilize analogs of UDP-galactose as donors (17). These include derivatives where OH-4 is epimerized or deoxygenated and where the 6-CH_2OH is replaced by hydrogen. We have recently also prepared the 3-deoxy-analog of UDP-galactose as summarized in Figure 7. UDP-3-deoxy-Gal (**13**) was found to a poor donor when βGlcNAc derivative **14**, was used as the acceptor (Eq. III). The rate of transfer measured with a coupled enzyme assay was only 0.16% that of the parent UDP-Gal. Despite this low activity, **14** (2 mg) could be completely converted to the 3'-deoxy-lactosamine disaccharide **15** in a 48 h incubation. The synthetic trisaccharide **16** was also an acceptor using UDP-3-deoxy-Gal as donor, allowing the preparation of the deoxytetrasaccharide **17** on a mg scale (Eq. IV). The purpose of preparing deoxy-disaccharide **15** was to evaluate its potential as an inhibitor for an $\alpha(2\rightarrow3)$ sialyltransferase. Similarly, deoxy-tetrasaccharide **17** was prepared in order to assess its inhibitory properties towards a $\beta(1\rightarrow3)$GlcNAc transferase, the so called "i" or extension enzyme involved in the biosynthesis of polylactosamines. Both deoxy-oligosaccharides **15** and **17** were found to be inactive.

$$\text{UDP-3-deoxy-Gal} + \beta\text{GlcNAc-O(CH}_2)_8\text{COOMe} \qquad (14)$$

(III) $\downarrow$ $\beta(1,4)$GalT

$$\text{3-deoxy-}\beta\text{Gal}(1\rightarrow4)\beta\text{GlcNAc-O(CH}_2)_8\text{COOMe} \qquad (15)$$

$$\text{UDP-3-deoxy-Gal} + \beta\text{GlcNAc}(1\rightarrow6)\alpha\text{Man}(1\rightarrow6)\beta\text{Man-O(CH}_2)_8\text{COOMe} \quad (16)$$

(IV) $\downarrow$ $\beta(1,4)$GalT

$$\text{3-deoxy-}\beta\text{Gal}(1\rightarrow4)\beta\text{GlcNAc}(1\rightarrow6)\alpha\text{Man}(1\rightarrow6)\beta\text{Man-O(CH}_2)_8\text{COOMe} \quad (17)$$

αMan(1→6)
                    βMan-O(CH$_2$)$_8$COOMe
αMan(1→3)

UDP-GlcNAc   GlcNAcT-I
Analogs   (human milk)
10-12

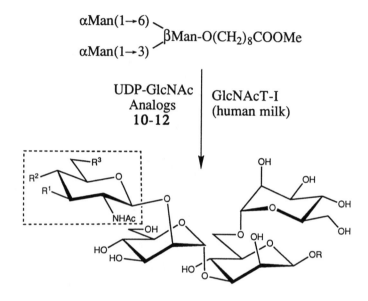

UDP-GlcNAc Analog		Product	%Yield
GlcNAc	→	R^1=OH, R^2=OH,R^3=OH	>97
10	→	R^1=H,R^2=OH,R^3=OH	70
11	→	R^1=OH,R^2=H,R^3=OH	10
12	→	R^1=OH,R^2=OH,R^3=H	35

Figure 5. Transfer of GlcNAc Analogs by GlcNAcT-I

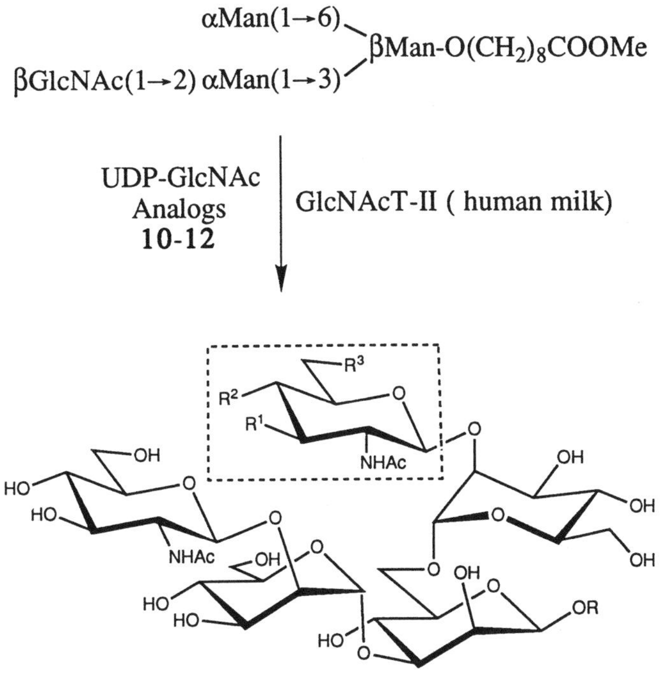

αMan(1$\rightarrow$6)
βGlcNAc(1$\rightarrow$2) αMan(1$\rightarrow$3)
βMan-O(CH$_2$)$_8$COOMe

UDP-GlcNAc
Analogs
10-12

GlcNAcT-II (human milk)

UDP-GlcNAc Analog		Product	%Yield
GlcNAc	$\rightarrow$	R^1=OH,R^2=OH,R^3=OH	100
10	$\rightarrow$	R^1=H,R^2=OH,R^3=OH	>95
11	$\rightarrow$	R^1=OH,R^2=H,R^3=OH	35
12	$\rightarrow$	R^1=OH,R^2=OH,R^3=H	>95

Figure 6. Transfer of GlcNAc Analogs by GlcNAcT-II.

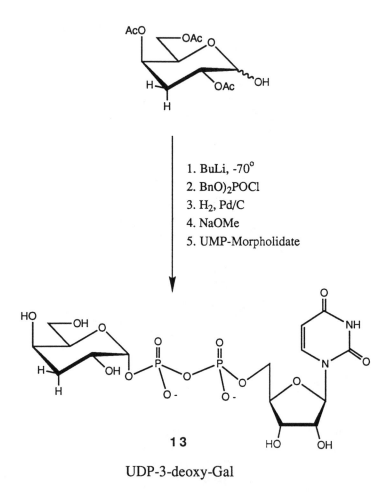

1. BuLi, -70°
2. BnO)$_2$POCl
3. H$_2$, Pd/C
4. NaOMe
5. UMP-Morpholidate

13

UDP-3-deoxy-Gal

Figure 7. Preparation of UDP-3-deoxy-Galactose.

In summary, all of the sugar-nucleotide analogs described above, as well as others previously reported in the literature, were sufficiently active as glycosyl-donors to allow the preparation of the corresponding oligosaccharide analogs on a mg scale. The reactions could undoubtedly be scaled up, but even on these small scales sufficient material was obtained to evaluate the products as enzyme-acceptors or as inhibitors. It is true that the chemical synthesis of sugar-nucleotide analogs is itself difficult and time consuming but the reader should note that only seven sugar-nucleotides are required for the biosynthesis of almost all mammalian glycoproteins and glycolipids. This means that only a quite limited number of sugar-nucleotides need to be prepared to allow the production of an essentially infinite number of oligosaccharide analogs. One reason we chose to investigate primarily the deoxygenated analogs of the sugar-nucleotides is that these are necessarily less sterically demanding than, for example, O-methylated or other derivatives. We are optimistic, however, that glycosyltransferases will become available, either by isolation or molecular cloning, that will be able to transfer glycosyl-residues larger than the parent monosaccharide, hopefully the size of disaccharides or larger. This optimism is based, in part, on the existence of bacterial glycosyltransferases which synthesize polysaccharides by the addition of preformed oligosaccharide blocks. Examples include the enzymes involved in the biosynthesis of cell-wall peptidoglycans (28) and capsular polysaccharides (29). Such glycosyltransferases could then be used in a "block-synthesis" of very large and complex oligosaccharides. Work towards this goal is in progress.

Acknowledgments

This research was supported by operating grants to OH and MMP from the Natural Sciences and Engineering Research Council of Canada.

Literature Cited

1. Watkins, W. M. Carbohydr. Res. 1986, 149, 1-12.
2. Bock, K.; Breimer, M. E.; Brignole, A.; Hansson, G. C.; Karlsson, K.-A.; Larson, G; Leffler, H; Samuelsson, B. E.; Stromberg, N.; Eden, C. S.; Thurin, J. J. Biol. Chem. 1985 260, 8545-8551.
3. Krivan, H. C.; Ginsburg, V; Roberts, D. D. Arch. Biochem. Biophys. (1988) 260, 493-496.
4. Krivan, H. C.; Roberts, D. D.; Ginsburg, V. Proc. Natl. Acad. Sci. USA 1988 85, 6157-6161.
5. Pritchett, T. J.; Brossmer, R.; Rose, U.; Paulson, J. C. Virology 1987 160, 502-506.
6. Hakomori, S,-I, Cancer Res. 1985 45, 2405-2414.
7. Fukuda, M. Biochem. Biophys. Acta 1985 780, 119-150.
8. Hakomori, S.-I. Annu. Rev. Immunol. 984 2, 103-126.
9. Feizi, T. Nature 1985 314, 53-57.
10. Paulsen, H. Angew. Chem. Int. Ed. Eng. 1982, 21, 155-173.
11. Ogawa, T.; Yamamoto, H.; Nukada, T.; Kitajima, T.;Sugimoto, M. Pure Appl. Chem. 1984, 56, 779-795.
12. Schmidt, R. R. Angew. Chem. Intl. Ed. Eng. 1986, 25, 212-235.
13. Toone, E. J.; Simon, E. S.; Bednarski, M. D.; Whitesides, G. M. Tetrahedron 1989, 45, 5365-5422.
14. Paulson, J.C.; Colley, K. J. J. Biol. Chem. 1989 264, 17615-17618.
15. Lemieux, R. U. Chem. Soc. Rev. 1989 18, 347-374.
16. Palcic, M. M.; Ripka, J.; Kaur, K. J.; Shoreibah, M.; Hindsgaul, O.; Pierce, M. J. Biol. Chem. 1990, 265, 6759-6769.

17. Berliner, L. J.; Robinson, R. D. Biochemistry 1982, 21, 6340-6343.
18. McDowell, W.; Grier, T. J.; Rasmussen, J. R.; Schwarz, R. T.
     Biochem. J. 1987, 248, 523-531.
19. Gross, H. J.; Bunsch, J. C.; Paulson, J. C.; Brossmer, R. Eur. J.
     Biochem. 1987, 168, 595-602.
20. Shibaev, V. N. Pure Appl. Chem. 1978, 50, 1421-1436.
21. Druzhinina, T. N.; Gogilashvili, L. M.; Shibaev, V. N. Bioorg. Khim.
     1988, 14, 1242-1249.
22. Palcic, M. M.; Venot, A. P.; Ratcliffe, R. M.; Hindsgaul, O.
     Carbohydr. Res. 1989, 190, 1-11.
23. Gokhale, U. B.; Hindsgaul, O.; Palcic, M. M. Canad. J. Chem. 1990,
     68, 1063-1071.
24. Schachter, H. Biochem. Cell. Biol. 1986, 64, 163-181.
25. Palcic, M. M.; Heerze, L. D.; Pierce, M.; Hindsgaul, O.
     Glycoconjugate J. 1988, 5, 49-63.
26. Srivastava, G.; Alton, G.; Hindsgaul, O. Carbohydr. Res. 1990, 207, 259-
     276.
27. Inage, M.; Chaki, H.; Kusumoto, S.; Shiba, T. Chem.Lett. 1982,
     1281-1284.
28. Ward, J. B. In Antibiotic Inhibitors of Bacterial Cell Wall Biosynthesis;
     Tipper, D. J. Ed; Pergamon Press; 1987, p.1-43.
29. Troy, F. A.; Ann. Rev. Microbiol., 1979, 33, 519-560.

RECEIVED December 4, 1990

Chapter 4

Use of Glycosidases and Glycosyltransferases in the Synthesis of Complex Oligosaccharides and Their Glycosides

K. G. I. Nilsson

Chemical Center, University of Lund, P.O. Box 124, S–22100 Lund, Sweden

This chapter describes recent syntheses of complex oligosaccharides using glycosidases and glycosyltransferases as catalysts. Glycosidases have been used for synthesis of several of the disaccharide sequences of glycoproteins and glycolipids. High yields were obtained by using high substrate concentrations. Thus, the yield of α-galactosidase-catalyzed synthesis of α-linked p-nitrophenyl digalactosides from p-nitrophenyl α-galactopyranoside was about 65% when the initial concentration of the substrate was 1 M, compared to a yield of only about 30% when the initial concentration of substrate was 0.15 M. The sequential use of glycosidases and glycosyltransferases was a convenient way to prepare sialylated trisaccharides. Both sialyltransferase and CMP-sialate synthase were immobilized to tresyl chloride-activated agarose in good yield (about 55%) and used for repeated synthesis with negligible loss of activity.

There is a growing interest in the production of the complex carbohydrate chains of glycoproteins and glycolipids due to the vital importance of these structures *in vivo* (1-4). The carbohydrate structures of glycoconjugates function as blood group determinants, are active as cell-surface receptors for proteins, toxins and pathogens, and are involved in cell-cell interactions (e.g. in organ development and lymphocyte migration). The importance of carbohydrate chains for the secretion, immunogenicity and circulation half-life of recombinant glycoproteins has been recognized (5).
   Fragments (disaccharides to pentasaccharides) of complex structures are often sufficient for full biological activity, of which some examples are listed in Table I (6). These structures have been produced mainly by chemical synthesis or by isolation from natural sources. Although these methods are well-developed for the preparation of mg- to g-quantites, they are not suitable for synthesis of g- to kg-quantities.

0097–6156/91/0466–0051$06.00/0
© 1991 American Chemical Society

Table I.   Examples of biologically active carbohydrates (*1-4*)

GalNAcα1-3(Fucα1-2)Gal	Blood group determinant A
Galα1-3(Fucα1-2)Gal	Blood group determinant B
GlcNAcβ1-3Gal	Receptor for *Streptococcus pneumoniae*
Galα1-4Gal	Receptor for uropathogenic *E. coli*
Galβ1-3GalNAc	Cancer-associated antigen (T-antigen)
Galβ1-4(Fucα1-3)GlcNAc	Cancer-associated antigen (Lewis-x)
NeuAcα2-3Galβ1-3GlcNAc	Cancer-associated antigen

The requirement for stereospecific reactions and the presence of multiple hydroxyl groups of similar and relatively low reactivity (*7, 8*) complicates the chemical synthesis of these oligosaccharides. Hazardous heavy metal salts (e.g. Ag-triflate, $Hg(CN)_2$) are often used as catalysts. To achieve good selectivity multiple protection and deprotection steps have to be carried out and the overall yields are often low. In particular, despite recent progress (*8*), the synthesis of important α-sialylated oligosaccharides is difficult: yields range from 20-30% and the unnatural β-sialoside is often formed and must be separated from the desired product.

Some years ago, we started a project aimed at the development of simple and efficient enzymatic methods for the synthesis of biologically-active carbohydrates on a preparative scale. The use of enzymes as catalysts for the synthesis of carbohydrates has many potential advantages, such as enabling selective, stereospecific syntheses with a minimum of reaction steps under mild conditions and in aqueous solutions in which carbohydrates are highly soluble. Moreover, many enzymes now can be produced in quantity by fermentation and can be immobilized and reused.

We have used two types of enzymes for the production of oligosaccharides: glycosidases (E.C. sub-class 3.2) and glycosyltransferases (E.C. sub-class 2.4) (*9, 10*). In this chapter we discuss the relative advantages and disadvantages of using each class of enzyme as a catalyst for the synthesis of complex carbohydrates.

Methods of Synthesis Using Glycosidases

Glycosidases have several advantages as catalysts compared with glycosyltransferases: they are present in relatively high concentrations in biological material and are easy to prepare; moreover, the substrates are easily available and can be used in high concentration.
Reversed Hydrolysis vs. Transglycosylation. Glycosidases have been used for several years for the selective hydrolysis of carbohydrates, such as for the hydrolysis of starch and for the analysis of glycoconjugate carbohydrate structures. These enzymes also can be used for the regioselective and stereospecific synthesis of oligosaccharides in equilibrium-controlled (reversed-hydrolysis, Equation 1) or kinetically-controlled (transglycosylation, Equation 2) reactions:

$$\text{Glucose + Glucose} \longrightarrow \text{Glucose-Glucose} + H_2O \qquad (1)$$

Lactose + Galactose $\longrightarrow$

$$\text{Lactose + Galactose} \left\{ \begin{array}{l} \longrightarrow \text{ Galactose-Galactose + Glucose} \\ \xrightarrow{\text{H}_2\text{O}} \text{ 2 Galactose + Glucose} \end{array} \right.$$

(2)

Equlibrium-controlled synthesis was described as early as 1898 by Hill (*11*) and transglycosylation with β-galactosidase was used in 1956 by Alessandrini *et al.* for the synthesis of Galβ1-3GlcNAc from lactose (glycosyl donor) and *N*-acetyl glucosamine (glycosyl acceptor) (*12*).

The primary advantages of the equilibrium-controlled approach (Equation 1) is the use of inexpensive, simple substrates. Good yields of disaccharides (30-40% yields of mixture of regioisomers, based on HPLC) have been obtained by using high substrate concentrations at increased reaction temperature (to increase the solubility of substrates and the rate of reactions) (*13, 14*). The separation of products is often difficult, however, because of the many regioisomers formed; isolated yields of products are often low (*13*). In a refinement of this approach, a carbon-celite column has been used as a "molecular trap" to enrich the product oligosaccharides (*15, 16*).

In transglycosylation reactions, the extent of oligosaccharide formation depends on the partition ratio of the glycosyl-enzyme intermediate between the transfer and hydrolytic reactions (Equation 2). Higher yields may be obtained than in equilibrium-controlled syntheses using the same concentration of acceptor. Moreover, transglycosylation reactions are more rapid than reversed-hydrolysis reactions due to the higher reactivity of the glycosyl donor. Oligosaccharides (e.g. lactose, Equation 2) or nitrophenyl glycosides have been frequently used as glycosyl donors in transglycosylation reactions (*9*).

Crude enzyme preparations can be used in transglycosylation reactions, but not in reversed-hydrolysis reactions. In transglycosylation reactions, contaminating glycosidases are of less importance, because transglycosylation usually is considerably faster than the competing equilibrium-controlled synthesis by the contaminating glycosidases. For example, a homogenate of bovine testes, which contained β-galactosidase, was used after centrifugation for the synthesis of various glycosides of β-linked Gal-GlcNAc from lactose and acceptor glycosides (*17*). Equilibrium-controlled reactions on the other hand, require pure enzyme preparations to minimize the formation of by-products. If, for example, the synthesis of Gal-GlcNAc is performed by reversed-hydrolysis from galactose and *N*-acetylglucosamine, it is important to use a pure β-galactosidase to minimize the possible synthesis of e.g. GlcNAc-Gal and α-linked Gal-GlcNAc by contaminating *N*-acetylglucosaminidase or α-galactosidase.

Synthesis of Disaccharides using Glycosidases

Regioselectivity of Glycosidase-Catalyzed Reactions. The primary disadvantage of using glycosidases rather than glycosyltransferases as catalysts is that reactions catalyzed by the former are not regiospecific and isolation of products is a complex task. Moreover, several glycosidases catalyze the preponderant formation of 1-6 linkages (*18, 19*), whereas most glycoconjugate structures of interest contain linkages to the secondary position of the acceptor, that is 1-2, 1-3, or 1-4 linkages. The problems of low and/or wrong regioselectivity has hindered the application of glycosidases for preparative synthesis of glycoconjugate structures, but, as shown below, these problems have now been largely solved for the synthesis of disaccharides.

Changing the regioselectivity of glycosidase-catalyzed synthesis with various acceptor glycosides. The regioselectivity of glycosidase-catalyzed transglycosylations can be conveniently manipulated and the isolation of products simplified by using glycosides, rather than carbohydrates with a free reducing end, as acceptors (e.g. Galβ-OR instead of galactose in Equation 2 where R is a suitable organic group) (20). Many disaccharide glycosides have been synthesized in preparative amounts (mg-kg scale) with regioselectivity from 70-95% and in yields of 10-65% using this method (Table II). The anomeric configuration and the aglycon structure of the acceptor glycoside influenced the regioselectivity of the reactions (20). For example, α-galactosidase from green coffee beans catalyzed mainly the formation of either α1-3, or α1-6-linked galactosides with Galα-OMe or Galβ-OMe, respectively, as acceptors. With β-galactosidase from *E. coli*, the methyl β-galactoside gave β1-3-linked digalactoside, whereas the α-glycoside gave the 1-6-linkage. Similar effects were observed with *N*-acetyl-α-galactosaminidase, *N*-acetyl-β-galactosaminidase and *N*-acetyl-β-glucosaminidase from *Chamelea gallina* (21).

The nature of the acceptor aglycon also influenced regioselectivity. Thus, with α-mannosidase from jack bean, the ratio of 1-2- and 1-6-linked products was 19:1 with *p*-nitrophenyl α-mannoside as acceptor, whereas with the corresponding methyl glycoside, the ratio was 5:1 (20). Similarly, with the bovine testes β-galactosidase, β1-3- and β1-4-linked Gal-GlcNAcβ-OMe was formed in about equal amounts with GlcNAcβ-OMe as acceptor, whereas the β1-3 isomer predominated (90%) with the trimethylsilylethyl glycoside as acceptor (Table II) (17).

Importance of the enzyme source. The source of the enzyme also affects regioselectivity. For example, whereas the *E. coli* β-galactosidase gave almost exclusive formation of β1-6-linked Gal-GlcNAc (26), the bovine testes enzyme gave a mixture of the β1-3, β1-4, and β1-6-linked isomers (12). Glycosidases from the mollusc *Chamelea gallina* catalyzed the synthesis of GalNAcα1-3Galα-OMe, GalNAcβ1-3Galβ-OMe, and GlcNAcβ1-3Galβ-OMe, whereas enzymes from several other sources were found to give predominant formation of 1-6-linkages (21).

Enzymatic hydrolysis of byproducts to control regioselectivity. In situations where the regioselectivity is low and several regioisomers are formed, the isolation of the desired product can be facilitated by the use of a second glycosidase with a different regioselectivity to selectively hydrolyze unwanted isomers. This approach has been used in the synthesis of Galβ1-3GalNAc and Galβ1-3GlcNAc with the bovine testes β-galactosidase (27). In this case, unwanted isomers were selectively hydrolyzed with the *E. coli* β-galactosidase after dilution of the reaction mixture.

Preparation of Acceptor Glycosides. The acceptor glycosides can be prepared enzymatically either in separate reactions or *in situ* as illustrated in Table II by the β-galactosidase catalyzed synthesis of mono- and disaccharide glycosides from lactose and alcohols (25). Although not

Table II.   Disaccharide   glycosides   synthesized   with   glycosidases

Glycosyl Donor	Glycosyl Acceptor	Major Glycoside Formed	Ref.
N-Acetyl-α-galactosaminidase			
GalNAcα-OPhNO$_2$-o	Galα-OMe	GalNAcα1-3Galα-OMe	21
N-Acetyl-β-galactosaminidase			
GalNAcβ-OPhNO$_2$-p	Galβ-OMe	GalNAcβ1-3Galβ-OMe	21
N-Acetyl-β-glucosaminidase			
GlcNAcβ-OPhNO$_2$-p	Galα-OMe	GlcNAcβ1-6Galα-OMe	21
GlcNAcβ-OPhNO$_2$-p	Galβ-OMe	GlcNAcβ1-3(6)Galβ-OMe	21
α-L-Fucosidase			
Fucα-OPhNO$_2$-p	Galβ-OMe	Fucα1-6Galβ-OMe	22
Fucα-OPhNO$_2$-p	GAlβ-OMe	Fucα1-2Galβ-OMe	23
α-Galactosidase			
Galα-OPhNO$_2$-o	Galα-OPhNO$_2$-o	Galα1-2Galα-OPhNO$_2$-o	20
Galα-OPhNO$_2$-p	GAlα-OPhNO$_2$-p	GAlα1-3Galα-OPhNO$_2$-p	20
Galα-OPhNO$_2$-p	Galα-OMe	Galα1-3Galα-OMe	20
Galα-OPhNO$_2$-p	Galα-OEtBr	Galα1-3Galα-OEtBr	24
Galα-OPhNO$_2$-p	Galβ-OMe	Galα1-6Galβ-OMe	20
Galα-OMe	Galα-OMe	Galα1-3Galα-OMe	24
Raffinose	Galα-OCH$_2$CH=CH$_2$	Galα1-3Galα-OCH$_2$CH=CH$_3$	25
Galα-OPhNO$_2$-p	GalNAcα-OEt	Galα1-3GalNAcα-OEt	17
β-Galactosidase			
Galβ-OPhNO$_2$-p	Galα-OMe	Galβ1-6Galα-OMe	20
Galβ-OPhNO$_2$-p	Galβ-OMe	Galβ1-3Galβ-OMe	20
Lactose	Allyl alcohol	Galβ-OCH$_2$CH=CH$_3$	25
		Galβ1-3Galβ-OCH$_2$CH=CH$_3$	
Lactose	Benzyl alcohol	Galβ-OPh	25
		Galβ1-3Galβ-OPh	
Lactose	Trimethylsilyl-ethanol	Galβ-OEtSiMe$_3$	25 Galβ-O
		Galβ1-3Galβ-OEtSiMe$_3$	
Galβ-OPhNO$_2$-p	GalNAcα-OEt	Galβ1-3GalNAcα-OEt	17
Galβ-OPhNO$_2$-p	GlcNAcβ-OMe	Galβ1-3(4)GlcNAcβ-OMe	17
Galβ-OPhNO$_2$-p	GlcNAcβ-OEtSiMe$_3$	Galβ1-3GlcNAcβ-OEtSiMe$_3$	17
α-Mannosidase			
Manα-OPhNO$_2$-p	Manα-OMe	Manα1-2Manα-OMe	20
Manα-OPhNO$_2$-p	Manα-OEtBr	Manα1-2Manα-OEtBr	20
Manα-OPhNO$_2$-p	Manα-OPhNO$_2$-p	Manα1-2Manα-OPhNO$_2$-p	20

optimized, about 20 g of allyl β-galactoside and 1 g of Galβ1-3Galβ-OCH2CH=CH2 was prepared in a one-pot reaction from lactose (100 g) and allyl-alcohol (50 ml) employing 2.5 mg of β-galactosidase (1550 U).

Isolation of Products. The isolation of products was drastically simplified by the use of acceptor glycosides instead of acceptors with a free reducing end since no anomerization of the product glycosides occurs. The products in Table II were isolated to homogeneity (usually above 99%) by column chromatography (silica, gel filtration) and characterized with standard methods (optical rotation, elemental analysis, melting point, permethylation analysis and NMR). Interestingly, the regioisomers of α-linked Gal-Gal-OPhNO2-p were separated completely in one step by gel filtration on Sephadex.

Use of Products. The glycosides shown in Table II can be used for a variety of purposes: allyl, benzyl and trimethylsilylethyl glycosides are useful for temporary anomeric protection; methyl glycosides can be used as inhibitors; and allyl, 2-bromoethyl or nitrophenyl glycosides can be used for the preparation of neoglycoconjugates or affinity adsorbents following further chemical modification. Thioethyl disaccharide glycosides are suitable as building blocks (after *O*-protection) for the chemical synthesis of higher oligosaccharides (7).

Yield of Glycosidase-Catalyzed Synthesis of Oligosaccharides. The yield of glycosidase-catalyzed synthesis depends on a number of parameters such as reaction time, temperature, organic solvent, and substrate concentration.

Effect of reaction time. Secondary hydrolysis of products during long reaction times will decrease the yield of transglycosylation (24). Different regioisomers may also be hydrolyzed at different rates and the regioselectivity of the reactions may, therefore, change during the course of reaction (24). It is important to monitor the progress of transglycosylation reactions so that the reaction may be stopped when the maximum yield is obtained.

Effect of temperature and organic solvents. The effect of these parameters on the yield of transglycosylations has not been extensively studied. In one study, it was found that the yield of α-galactosidase-catalyzed synthesis of α-linked *p*-nitrophenyl digalactosides increased from 32 to 47% by using a lower reaction temperature (4°C instead of 50°C) (24). The regioselectivity was also improved from 77 to 90%. Addition of organic cosolvent was also found to decrease the yield of this reaction from 32% in buffer to 10% in 45% *N*,*N*-dimethylformamide. This effect was attributed to a decreased hydrophobic interaction between enzyme and acceptor upon addition of the cosolvent (28). A similar effect was observed in the synthesis of *p*-nitrophenyl dimannosides from Mana-OPhNO2-p using α-mannosidase (Nilsson, K. G. I., unpublished data). In another study with lysozyme, however, the yield of (GlcNAc)5-OPhNO2-p was found to increase upon addition of up to 60% methanol (29).

Carbohydrates have a low solubility in organic solvents and there are few reports on the behavior of glycosidases in almost 100% organic solvent. Bourquelot and Bridel described the synthesis of various alkyl β-glucosides from glucose and alcohols with the enzyme suspended in high concentrations of the respective alcohol (60-95%, v/v) (30). The yield of ethyl β-glucoside was reported to be about 90% in 95% (v/v) ethanol.

It was found that α-mannosidase (jack bean, 10 units/mL, Sigma) was active in high concentrations of organic solvent (Nilsson, K. G. I., unpublished data). Thus, the enzyme catalyzed the formation of p-nitrophenyl dimannosides from Manα-OPhNO$_2$-p in 97% tetrahydrofuran with the concomitant release of p-nitrophenol. The rate of reaction, however, was very slow compared to the rate in buffer. The reaction was followed by measuring the amount of nitrophenol released (UV, 405 nm) and with HPLC (reversed-phase chromatography).

    Effect of substrate concentration. The yields of transglycosylations have generally been in the range 10-50%, which is lower than the yields reported with glycosyltransferases (30-95%). Water competes with the acceptor nucleophile for the glycosyl-enzyme intermediate in transglycosylation reactions (Equation 2). An obvious way to decrease the hydrolytic side-reaction is to increase the concentration of substrates. In previous studies, substrate concentrations far from saturation have been used.

    It was found that the yields of α-galactosidase-catalyzed formation of α-linked Gal-Galα-OPhNO$_2$-p increased from 30 to 65% when the initial concentration of the substrate Galα-OPhNO$_2$-p was increased from 0.15 to 1 M (Nilsson, K. G. I., unpublished data). The total yield of α-linked digalactosides was about 75% when higher substrate concentrations. The reaction was carried out at elevated temperature (50 °C) to avoid substrate precipitation. Similarly, the relatively low yield (10%) of β-linked GlcNAc-Galβ-OMe in the N-acetyl-β-glucosaminidase-catalyzed transglycosylation of GlcNAcβ-OPhNO$_2$-p with 0.5 M Galβ-OMe was considerably improved by increasing the acceptor concentration (Table III) (*21*). In another example

Table III.   Yield of Glycosidase-catalyzed Disaccharide Formation at
Various Substrate Concentrations

Concentration of Acceptor (%; w/w)	Yield (%) of	
	Gal α–Gal α-OPhNO$_2$-p	GlcNAcβ-Galβ-OMe
4.5	30	n.d.
10	n.d.	10
22	65	22
40	n.d.	42

the acceptor glycoside GlcNAcβ-OEtSiMe$_3$, which has a low solubility in water, was converted in good yield (55%) to product glycoside (Galβ1-3GlcNAcβ-OEtSiMe$_3$), with the efficient donor Galβ-OPhNO$_2$-p, which was added in excess to minimize secondary hydrolysis (*17*).

Synthesis of Higher Oligosaccharides using Glycosidases

Use of Exoglycosidases. Some tri- and tetrasaccharide structures present in glycoconjugates have been prepared with exoglycosidases. The methyl glycoside of the structure Manα1-2Manα1-2Man, which is a common sequence of glycoprotein high-mannose glycans, accumulated together with other regioisomers in the later stages of the α-mannosidase-catalyzed synthesis of Manα1-2Manα-OMe from Manα-OPhNO$_2$-p (Manα-OR) and Manα-OMe (20):

$$\text{Man}\alpha\text{-OR} + \text{Man}\alpha\text{-OMe} \xrightarrow{-\text{HOR}} \text{Man}\alpha1\text{-}2\text{Man}\alpha\text{-OMe} \qquad (3)$$

$$\text{Man}\alpha\text{-OR} + \text{Man}\alpha1\text{-}2\text{Man}\alpha\text{-OMe} \xrightarrow{-\text{HOR}} \text{Man}\alpha1\text{-}2\text{Man}\alpha1\text{-}2\text{Man}\alpha\text{-OMe} \qquad (4)$$

Galβ1-3Galβ1-4GlcNAcβ-OEtSiMe₃ was formed from Galβ-OPhNO₂-p and GlcNAcβ-OEtSiMe₃ in a similar reaction with β-galactosidase from bovine testes as catalyst (17). This reaction showed good regio- and acceptor selectivity (the major disaccharide formed, Galβ1-3GlcNAcβ-OEtSiMe₃, was a poor acceptor).

The cancer-associated structure Galα1-3Galβ1-4GlcNAcβ-OEt was formed by using β-galactosidase and α-galactosidase in sequence (31). Similarly, lacto-N-tetraose, Galβ1-3GlcNAcβ1-3Galβ1-4Glc, was prepared by the sequential use of N-acetyl-β-glucosaminidase and β-galactosidase (Nilsson, K. G. I. et al., unpublished data):

$$\text{GlcNAc}\beta\text{-OR} + \text{lactose} \xrightarrow{-\text{HOR}} \text{GlcNAc}\beta1\text{-}3\text{Gal}\beta1\text{-}4\text{Glc} \qquad (5)$$

$$\text{Lactose} + \text{GlcNAc}\beta1\text{-}3\text{Gal}\beta1\text{-}4\text{Glc} \xrightarrow{-\text{Glc}} \text{Gal}\beta1\text{-}3\text{Gal}\beta1\text{-}3\text{Gal}\beta1\text{-}4\text{Glc} \qquad (6)$$

The first reaction was carried out either as an equilibrium reaction (R=H; yield ca. 3%) or as a transglycosylation reaction (R=PhNO₂-p; yield ca. 10%). Several regioisomers were formed in the reactions.

Use of Endoglycosidases. The yields and regioselectivities of the above exoglycosidase-catalyzed reactions are generally low. Endo-α-N-acetylgalactosaminidase from *Diplococcus pneumoniae* (32) and endo-β-N-acetylglucosaminidase F (33) have been shown to have transfer activity. The use of endoglycosidases for synthesis of higher oligosaccharides remains to be investigated. One approach would be to use exoglycosidases for synthesis of shorter saccharide units and endoglycosidases to join the fragments. Some endoglycosidases may be useful for *in vitro* synthesis of glycolipids and glycopeptides.

Sequential Use of Glycosidases and Glycosyltransferases

Glycosyltransferases are considered to be responsible for the specific synthesis of the various glycoconjugate glycans *in vivo* (34). The following equation represents the general reaction catalyzed by glycosyltransferases:

$$\text{NDP-Donor} + \text{Acceptor} \longrightarrow \text{Donor-Acceptor} + \text{NDP} \qquad (7)$$

where NDP-Donor is a nucleoside diphosphate sugar having N being either U (uridine) or G (guanosine), depending on the type of glycosyltransferase. The sialyltransferases use cytidine 5'-N-acetylneuraminic acid, CMP-NeuAc, as a donor. The advantage of glycosyltransferases compared with glycosidases as catalysts for the synthesis of carbohydrates is that they form only one regioisomer from a given glycosyl donor (nucleoside-sugar) and acceptor mono- or oligosaccharide. In many cases the enzymes are highly specific for the acceptor and so it is possible to glycosylate only one sugar or type of sugar present either as a component of a mixture of carbohydrates or as part of a larger oligosaccharide structure.

The present disadvantage of using glycosyltransferases as catalysts is the inavailability of most enzymes, although some glycosyltransferases have been cloned and in a few years several more may be commercially available (*35-37*). Moreover, substrate/product inhibition has necessitated the use of low substrate concentrations (*38*) in some cases.

Several oligosaccharide structures have been prepared *in vitro* in good yield (30-90%) with different glycosyltransferases (α1-2fucosyl-, β1-4galactosyl-, α2-3sialyl-, and α2-6sialyltransferases) (*9, 10, 17, 38-47*). In general, the higher oligosaccharide structures were prepared by means of a chemoenzymatic approach: chemical methods were used for the synthesis of acceptor oligosaccharides and glycosyltransferases were used for the final glycosylation steps. In a alternative approach, abundant glycosidases are used to produce shorter oligosaccharides and glycosyltransferases are used for the final glycosylation steps where the demand on regiospecificity is high (*17*). This approach is exemplified below.

Synthesis of Sialyated Trisaccharides. We have used a combination of glycosidases and glycosyltranferases to prepare sialylated trisaccharides (Equations 8-9). The enzyme β-galactosidase from bovine

$$\text{Galβ-OPhNO}_2\text{-p+GalNAcα-OEt} \xrightarrow{\text{-HOPhNO}_2\text{-p}} \text{Galβ1-3GalNAcα-OEt} \quad (8)$$

$$\text{CMP-NeuAc+Galβ1-3GalNAcα-OEt} \xrightarrow{\text{-CMP}} \text{NeuAcα2-3Galβ1-3GalNAcα-OEt} (9)$$

testes was used in the first reaction and an α2-3sialyltransferase from porcine submaxillary gland in the second reaction to give a 72% isolated yield of sialylated product (*17*). The sialyltransferase reaction has a preference for acceptors of the type Galβ1-3GalNAc, but NeuAcα2-3Galβ1-3cNAcβ-OR (R=methyl or 2-bromoethyl) was also obtained in good yield (30% isolated yield).

Isolation and immobilization of sialyltransferases. The porcine sialyltransferase is commercially available but is prohibitively expensive. The enzyme can also be isolated by affinity chromatography, provided a suitable affinity adsorbent is available such as CDP-hexylamine-agarose (*34*). We found that both a commercially-available preparation and a partially-purified preparation (obtained after homogenization, extraction, ion-exchange chromatography and one affinity chromatographic step) could be immobilized under mild conditions (pH 7.5, room temperature, 2 h) to tresyl chloride activated agarose with high retention of added activity (80% yield for commercially-available preparation, and 55% yield for partially-purified preparation) (*48, 49*).

The immobilized enzyme was used for repeated synthesis of the above sialylated trisaccharides without adding detergent to the system, which was used in previous synthesis with the soluble enzyme (*42*). The use of detergent complicates isolation of the products. The stability of the enzyme preparation was good and no decrease of activity was observed on repeated preparative synthesis. Recently, immobilization to CNBr-activated agarose of the α2-6sialyltransferase was reported (*50*).

Synthesis of CMP-NeuAc. CMP-NeuAc can be prepared from CTP and NeuAc using CMP-NeuAc synthase (Equation 10). Several reports on the enzymatic synthesis of nucleotide-sugars have appeared (*10*) and CMP-NeuAc synthase have been cloned (*51*). A conveniently obtained immobi-

$$CTP + NeuAc \xrightarrow{\text{CMP-NeuAc synthase}} CMP\text{-}NeuAc + PPi \qquad (10)$$

lized synthase preparation was found to be suitable for the preparative synthesis of CMP-NeuAc (48, 49). The immobilized preparation was obtained by reacting a crude extract of bovine liver with tresyl chloride-activated agarose under mild conditions (2 h, room temperature, pH 8). Immobilization eliminated the problem of contamination of the reaction mixture with the various low-molecular weight substances present in crude extracts and the reaction could be followed using FPLC (49).

We found that the enzyme preparation converted N-acetylneuraminic acid almost completely to CMP-NeuAc (90% yield, FPLC) at pH 9 and with the starting concentrations of 10 mM of NeuAc and 15 mM of CTP. Moreover, the enzyme preparation showed only a slight decrease of activity after three separate uses for a total reaction time of 90 h at 30°C (49).

Conclusions

The production of active glycosyltransferases with recombinant techniques and the development of efficient methods for the production of nucleoside-sugars are important for the application of glycosyltransferases in the synthesis of oligosaccharides on a large scale. The simple reaction systems that can be used with glycosidases, their availability as well as their tolerance to a wide range of acceptor structures make these catalysts highly attractive for synthesis of complex oligosaccharides and their analogs on a larger scale. High yields can be obtained with glycosidases using high substrate concentrations. The use of endoglycosidases may be instrumental for the synthesis of higher oligosaccharides.

Literature Citied

1.   Ginsburg, V.; Robbins, P. W. Biology of Carbohydrates; Ginsburg, V.; Robbins, P. W., Eds; Wiley: New York, 1984; Vol. 2.
2.   Chemistry and Physics of Lipids; Brady, R., Ed.; Elsevier: New York, 1986; p 42.
3.   Hakomori, S. Ann. Rev. Immunol. 1984, 2, 103-126.
4.   Karlsson, K. -A. Ann. Rev. Biochem. 1989, 58, 309-350.
5.   Tsuda, E.; Goto, M.; Murkami, A.; Akai, K.; Ueda, M.; Kawanishi, G.; Takahashi, N.; Sasaki, R.; Chiba, H.; Ishihara, H.; Mori, M.; Tejima, S.; Endo, S.; Arata, Y. Biochemistry 1988, 27, 5646-5654.
6.   Abbreviated Nomenclature According to IUB-IUPAC Recommendations: J. Biol. Chem. 1982, 257, 3347-3351.
7.   Schmidt, R. R. Angew. Chem. 1986, 98, 213-236.
8.   Marra, A.; Sinay, P. Carbohydr. Res. 1990, 195, 303-308.
9.   Nilsson, K. G. I. Trends Biotechnol. 1988, 6, 256-264.
10. Toone, E. J.; Simon, E. S.; Bednarski, M. D.; Whitesides, G. M. Tetrahedron 1989, 45, 5365-5422.
11. Hill, A. C. J. Chem. Soc. 1898, 73, 634-658.
12. Alsesandrini, A.; Schmidt, E.; Zilliken, F.; György, P. J. Biol. Chem. 1956, 220, 71-78.

13. Johansson, E.; Hedbys, L.; Mosbach, K.; Larsson, P. -O.; Gunnarsson, A.; Svensson, S. *Biotechnol. Lett.* **1986**, *8*, 421-424.
14. Ajisaka, K.; Nishida, H.; Fujimoto, H. *Biotechnol. Lett.* **1987**, *9*, 243-248.
15. Wallenfels, K. *Bull. Soc. Chim. Biol.* **1960**, *42*, 1715-1729.
16. Ajisaka, K.; Nishida, H.; Fujimoto, H. *Biotechnol. Lett.* **1987**, *9*, 387-392.
17. Nilsson, K. G. I. *Carbohydr. Res.* **1989**, *188*, 9-17.
18. Dey, P. M.; Pridham, J. B. *Adv. Enzymol.* **1972**, *36*, 91-130.
19. Wallenfels, K.; Weil, R. *The Enzymes*, **1972**, *7*, 617-663.
20. Nilsson, K. G. I. *Carbohydr. Res.* **1987**, *167*, 95-103.
21. Nilsson, K. G. I. *Carbohydr. Res.* **1990**, *204*, 79-83.
22. Nilsson, K. G. I. In *Opportunities in Biotransformation*; Copping, L. G.; Martin, R. E.; Pickett, Z. A.; Bucke, C.; Bunch, A. W., Eds.; Elsevier: Amsterdam, 1990; pp 131-139.
23. Svensson, S. C. T.; Thiem, J. *Carbohydr. Res.* **1990**, *200*, 391-402.
24. Nilsson, K. G. I. *Ann. N.Y. Acad. Sci.* **1988**, *542*, 383-389.
25. Nilsson, K. G. I. *Carbohydr. Res.* **1988**, *180*, 53-59.
26. Kuhn, R.; Baer, H. H.; Gauhe, A. *Chem. Ber.* **1955**, *188*, 1713-1723.
27. Hedbys, L.; Johansson, E.; Mosbach, K.; Larsson, P. -O.; Gunnarsson, A.; Svensson, S.; Lönn, H. *Glycoconjugate J.* **1989**, *6*, 161-168.
28. Nilsson, K. G. I. In *Studies in Organic Chemistry*; Laane, C.; Tramper, J.; Lilly, M. D., Eds.; Elsevier: Amsterdam, 1990; Vol. 29, pp 369-374.
29. Usui, T.; Hayashi, Y.; Nanjo, F.; Ishido, Y. *Biochem. Biophys. Acta* **1988**, *953*, 179-184.
30. Bourquelot, E.; Bridel, M. *Ann. Chim.* **1913**, *29*, 145-218.
31. Nilsson, K. G. I., Proc. 5th European Symp. Carbohydr.; Prague, **1989**, August 21-25, pp C-76.
32. Bardales, R. M.; Bhavanandan, V. P. *J. Biol. Chem.* **1989**, *264*, 19893-19897.
33. Trimble, R. B.; Atkinson, P. H.; Tarentino; A. L.; Plummer, T.; Maley, F.; Tomer, K. B. *J. Biol. Chem.* **1986**, *261*, 12000-12005.
34. Beyer, T. A.; Sadler, J. E.; Rearick, J. I.; Paulson, J. C.; Hill, R. L. *Adv. Enzymol.* **1981**, *52*, 23-176.
35. Rajan, V. P.; Larsen, R. D.; Ajmeras, S.; Ernst, L. K.; Lowe, J. B. *J. Biol. Chem.* **1989**, *264*, 11158-11167.
36. Joziasse, D. H.; Shaper, J. H.; van den Eijnden, D. H.; van Tunen, A. J.; Shaper, N. L. *J. Biol. Chem.* **1989**, *264*, 14290-14297.
37. Colley, K. J.; Lee, E. U.; Adler, B.; Browne, J. K.; Paulson, J. C. *J. Biol. Chem.* **1989**, *264*, 17619-17622.
38. Nunez, H. A.; Barker, R. *Biochemistry* **1980**, *19*, 489-495.
39. Rosevear, P. R.; Nunez, H. A.; Barker, R. *Biochemistry* **1982**, *21*, 1421-1431.
40. Wong, S. -H.; Haynie, S. L.; Whitesides, G. M. *J. Org. Chem.* **1982**, *47*, 5416-5418.
41. Auge, C.; David, S.; Mathieu, C.; Gautheron, C. *Tetrahedron Lett.* **1984**, *25*, 1467-1470.
42. Sabesan, S.; Paulson, J. C. *J. Am. Chem. Soc.* **1986**, *108*, 2068-2080.
43. Thiem, J.; Treder, W. F. *Ang. Chem.* **1986**, *98*, 1100-1101.
44. Nilsson, K. G. I., Proc. Sialic Acids **1988**, Schauer, R.; Yamakawa, T., Eds.; Berlin, 28-29.
45. de Heij, H. T.; Kloosterman, M.; Koppen, P.L.; van den Boom, J. H.; van den Eijnden, D. H. *J. Carbohydr. Chem.* **1988**, *7*, 209-222.
46. Crawley, S. C.; Hindsgaul, O.; Ratcliffe, R. M.; Lamontagne, L. R.; Palcic, M. M. *Carbohydr. Res.* **1989**, *193*, 249-256.

47. Palcic, M. M.; Venot, A. P.; Ratcliffe, R. M.; Hindsgaul, O. *Carbohydr. Res.* **1989**, *190*, 1-11.
48. Nilsson, K. G. I.; Gudmundsson, B. -M. E. Proc. Sialic Acids **1988**, Schauer, R.; Yamakawa, T., Eds.; Berlin, 30-31.
49. Nilsson, K.G.I. and Gudmundsson, B. -M. E. *Meth. Mol. Cellul. Biol.* **1990**, 1, 195-202.
50. Augé C.; Fernandez-Fernandez, R.; Gautheron, C. *Carbohydr. Res.* **1990**, *200*, 257-268.
51. Zapata, G.; Vann, W. F.; Aaronsson, W.; Lewis, M. S.; Moos, M. *J. Biol. Chem.* **1989**, *264*, 14769-14774.

RECEIVED February 1, 1991

Chapter 5

Glycosylation by Use of Glycohydrolases and Glycosyltransferases in Preparative Scale

Peter Stangier and Joachim Thiem

Institut für Organische Chemie, Universität Hamburg,
Martin-Luther-King-Platz 6, D–2000 Hamburg 13, Germany

Preparation of glycosides by use of enzymes is described. Hydrolases are applied for fucosylation and sialylation of model compounds in up to 20% yield. Glucosylation and galactosylation can be achieved to give di- and trisaccharides starting with unnatural glycosyl fluorides. The enzymatic regeneration of the cofactor CTP can be combined with its use in the enzymatic synthesis of activated neuraminic acid (CMP-Neu5Ac). This derivative is applied in the trisaccharide synthesis of Neu5Acα(2-6)Galβ(1-4)GlcNAc. Further, the activated fucose (GDP-Fuc) is obtained by a preparative enzymatic process. Finally, a series of galactosylations are successfully performed using galactosyl transferase and various glycoprotein models of lower saccharide amino acid conjugates.

Within the recent years, a variety of complex biological processes have been unraveled which proved glycoproteins (1) and glycolipids (2) to be carriers of significant biological information and to be essential for intercellular recognition, differentiation and growth (3). The specific shape of such complex oligosaccharide protein structures located on the extracellular side of membrane-bound peptides is thought to be responsible for these effects (Scheme 1).

The high variability of oligosaccharide structures originates from the highly different sequences of saccharide monomers, their mode of linkage, or anomeric configuration, and site of branching. Thus, classical organic synthesis, in particular, in this field does not only require an elaborate technique in the application of protective group chemistry and the patterns thereof, but also a profound knowledge in the field of glycosylation processes (4). Altogether, generally, this results in lengthy procedures with rather low overall yields. The majority of such syntheses were achieved by only a handful of very specialized groups in the world.

In recent years, in addition to the large number of classical contributions to glycosylations which have been sought, the most promising attempts for the time being have been proposed to apply

0097–6156/91/0466–0063$06.00/0

enzymatic routes by copying nature's approaches both for construction and cleavage of such complex systems. Taking advantage of chiral catalysts, namely enzymes, the principle of protection of hydroxyl groups can be skipped totally while the enzyme itself serves as a temporary and intermediate protecting group.

Principally, there are two entirely different groups of enzymes utilized in such syntheses: glycosyltransferases (5) and glycohydrolases (6) (Scheme 2). The former specifically builds up interglycosidic linkages from nucleotide sugars. The latter naturally hydrolyses the interglycosidic linkage, but by reversion of their natural properties under certain optimized reaction conditions, may be used for synthetic purposes. Glycohydrolyases do not require difficult isolation procedures, nor cofactors. The drawback, however, is poor regioselectivity and, generally, lower yields.

Application of Glycohydrolases

The biological function (7) of the hydrolase α-L-fucosidase (E.C. 3.2.1.51) is the hydrolysis of L-configurated fucosides (Scheme 3). α-L-Fucosidases from bacteria and moulds, mammals and molluscs are classified broadly into three groups based on their aglycon specificity (8). In humans, the genetically-linked deficiency of this enzyme results in the neurovisceral storage disease, fucosidasis (9).

Interest in fucosylation of glycoconjugates is connected with higher fucosylation rates in aberrant tissue. Thus it was attempted to use enzymatic fucosylation processes to obtain fucosylated oligosaccharides. The isolation of α-L-fucosidase from porcine liver is performed, after ammonium sulfate precipitation, by affinity chromatography with both CNBr Sepharose-bound β-L-fucopyranosyl amine 1 and (3-amino)propyl α-L-fucopyranoside 2 (10) (Scheme 4). The latter, comprising a new "C-glycoside" is synthesized via a radical reaction (11) starting from peracetylated fucosyl bromide (12) and acrylonitrile. The protein elution profile clearly demonstrates the acceptance of 2 by the enzyme, resulting in a activity yield of 78% with a purification factor of 3770. The specific activity of the eluate is 15 U/mg. Even though the enzyme is thought to be specific for the a-fucosidic linkage (cf. Scheme 3) it surprisingly accepts β-fucopyranosyl amine 1 as an affinity ligand to result, after an affinity chromatographic step in yields of 75%, with a purification factor of approximately 3400 and a specific activity of 13.6 U/mg.

The enzymatic glycosylation is carried out with both donors, p-nitrophenyl α-L-fucopyranoside 3 and α-L-fucopyranosyl fluoride 4, transferred onto the acceptor methyl β-D-galactopyranoside 5 (10) (Scheme 5). The two products obtained in this reaction sequence in 16.5% overall yield, L-Fucα(1-2)Galβ(1-OMe) 6 and L-Fucα(1-6)Galβ(1-OMe) 7, are found to be either α(1-2)- or α(1-6)- linked in a ratio of 1:2, respectively (cf. Table 1).

Due to the poor solubility of 3 in water either N, N'-dimethylformamide, dimethyl sulfoxide or acetonitrile were applied as cosolvents, all of which, however, cause in a drop of enzyme activity by approximately 50%. In further experiments the fucosidase is immobilized on Sepharose 4B as solid support. Transfer of L-fucose is observed to give the same product distribution as discussed earlier.

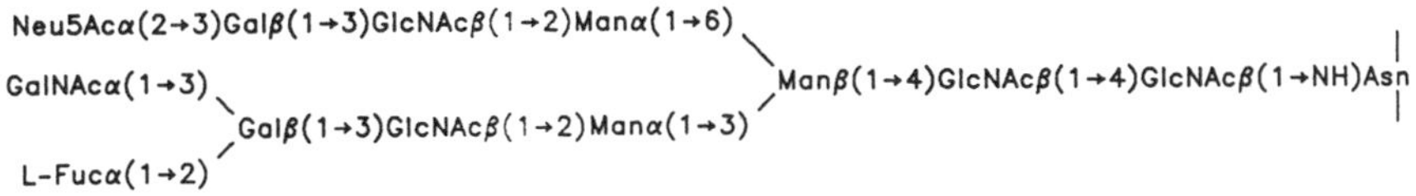

Neu5Acα(2→3)Galβ(1→3)GlcNAcβ(1→2)Manα(1→6)

GalNAcα(1→3)

Galβ(1→3)GlcNAcβ(1→2)Manα(1→3)

L-Fucα(1→2)

Manβ(1→4)GlcNAcβ(1→4)GlcNAcβ(1→NH)Asn

Scheme 1.   Extracellular oligosaccharide structure.

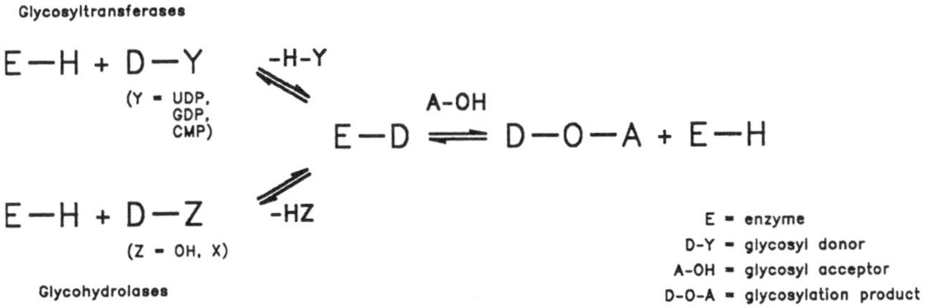

Glycosyltransferases

E—H + D—Y ⇌ -H-Y

(Y = UDP, GDP, CMP)

A-OH

E—D ⇌ D—O—A + E—H

E—H + D—Z ⇌ -HZ

(Z = OH, X)

Glycohydrolases

E = enzyme
D-Y = glycosyl donor
A-OH = glycosyl acceptor
D-O-A = glycosylation product

Scheme 2.   Formation of interglycosidic linkages by enzymatic processes.

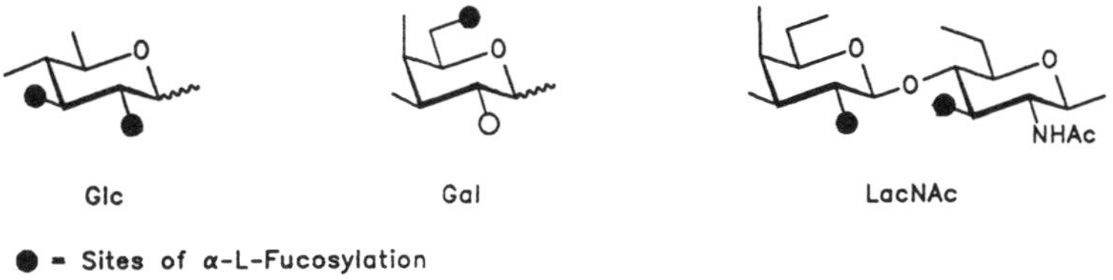

Glc   Gal   LacNAc   NHAc

● = Sites of α-L-Fucosylation

Scheme 3.   Naturally occurring sites of α-L-fucosylation.

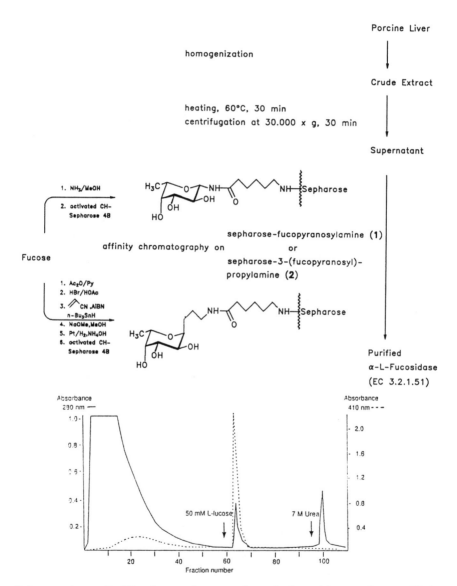

Scheme 4. Purification scheme of α-L-fucosidase by affinity chromatography and elution profile of enzyme.

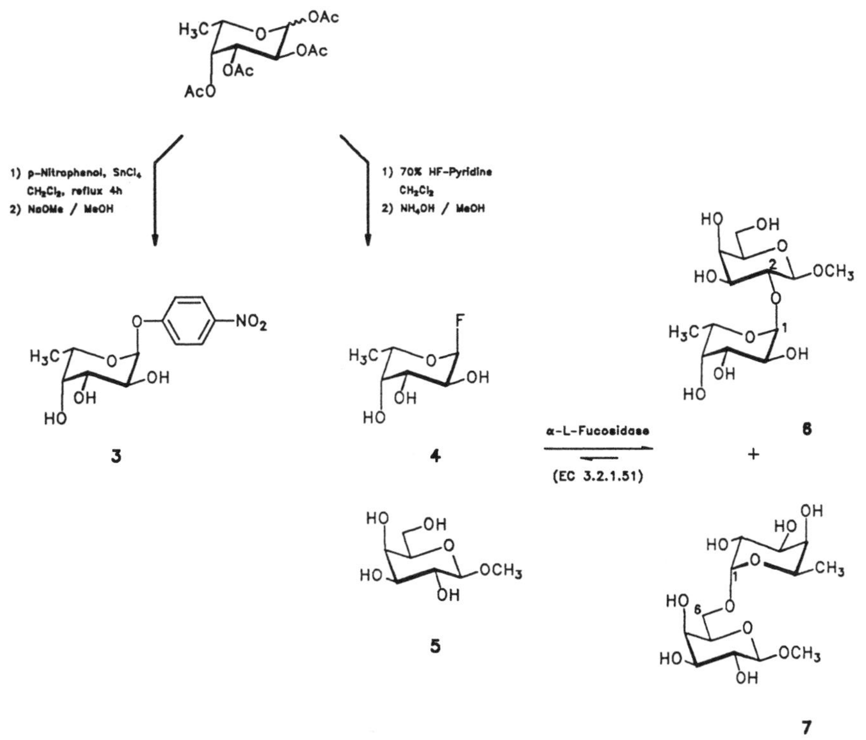

Scheme 5.   Syntheses of fucose derivatives and enzymatic formation of disaccharides by α-L-fucosidase.

Table 1. Enzymatic glycosylation using porcine α-L-fucosidase[a]

| Glycosyl Donor (1mmol) | Conditions 0.1 M NaH$_2$PO$_4$ buffer, pH 4.8 | Reaction time | Yield (%)[b] | | |
			L-Fucα(1→2) Galβ(1→OMe) (6)	L-Fucα(1→6) Galβ(1→OMe) (7)	Total
FucF(4)	Pure buffer	7 h	2.9	6.1	9.0
FucF(4)	30 % DMSO	24 h	5.0	4.7	9.7
pNPFuc(3)	30 % DMSO	5 d	6.5	10.0	16.5
pNPFuc(3)	Immobil.enzyme[c]	24 h	2.5	6.5	8.5

[a] 2 U of α-L-fucosidase in 2.1 ml buffer solvent containing 2.5 mmol methyl α-D-galactopyranoside (5) as acceptor; stirring at 37°C
[b] Yields of peracetylated compounds.
[c] Immobilized enzyme (4U, activ. Sepharose 4B) in 5 ml buffer, stirring at room temperature.

Another hydrolase used advantageously by our group for transglycosylation is the neuraminidase from *Vibrio cholerae* (*13*) (E.C. 3.2.1.18). By reversal of its physiological properties, the hydrolysis of the Neu5Acα(2-3)Gal and Neu5Acα(2-6)Gal interglycosidic linkages (Scheme 6), is forced towards its build up. This enzyme, therefore, is of major interest in the synthesis of neuraminic acid oligosaccharides due to their important role in intercellular processes, virus interactions, and cell differentiation (*14*).

Again, the acceptor is methyl α-D-galactopyranoside **5**, which is reacted with *p*-nitrophenyl-*N*-acetyl neuraminic acid **8**. The product Neu5Acα(2-3)Galβ(1-OMe) **9** and Neu5Acα(2-6)Galβ(1-OMe) **10** are formed in up to 9% total yield, respectively (*15*) (Table 2). Formation of the α(2-6) linked compound **10** is favored. This may be rationalized either by the assumptions of a higher nucleophilicity of the primary hydroxyl group or by the favored cleavage of the α(2-3) glycosidic bond as the reverse reaction. Similar results with yields up to 15% were observed for neuraminidase immobilized on VA-epoxy as solid support.

Both reactions using α-glucosidase (E.C. 3.2.1.20) or α-galactosidase (E.C. 3.2.1.22) are carried out with α-glycosyl fluorides of *gluco* and *galacto* configuration, respectively, both as donor and acceptor systems (*16*) (Scheme 7). α-Glucopyranosylfluoride **11** gives α-isomaltosyl fluoride in 25% yield **12** and 6% of isomaltose **13** and 6% of panose **14**, when incubated in high concentration with α-glucosidase. In other experiments with α-galactosyl fluoride **15**, α-galactosidase Galα(1-6)Galα(1-F) **16** is formed in 9% yield.

Enzymatic Syntheses with Glycosyltransferases

Activated glycosyl donors for enzymatic glycosylation of nucleotide diphosphosugars, are synthesized from glycosyl phosphates and nucleoside triphosphates. In 1950, uridine 5'-diphosphoglucose (UDP-Glc) was characterized as the first member of this class of activated glycosyl donors (*17*). This could then be shown to be in common with all other naturally occurring nucleotide sugars (*18*). In mammalian cells, glucose, galactose, *N*-acetylglucosamine, *N*-acetylgalactosamine, xylose, and glucuronic acid are activated via their uridine 5'-diphosphate (UDP) derivatives, whereas, mannose and L-fucose are present as guanosine 5'-diphosphate (GDP) derivatives. Unlike the diphosphosugars mentioned, sialic acids are found to be anomerically linked via a single phosphate unit to the base cytidine.

In order to utilize the high synthetic potential of these transferases, the nucleotide triphosphates have to be available. Fermentation and chemical synthesis were the methods at hand until recently, when the more advantageous enzymatic approaches have been devised, as it is exemplified by the regeneration of CTP (*19, 20*) or GTP (Scheme 8). Starting with the corresponding monophosphates cytidine 5'-monophosphate **17** or guanosine 5'-monophosphate **18** the products cytidine 5'-diphosphate **19** or guanosine 5'-diphosphate **20** are formed in the presence of adenylate kinase (E.C. 2.7.4.3) or guanylate kinase (E.C. 2.7.4.8), respectively. The first phosphate unit added, is transferred from adenosine 5'-triphosphate **21**, the donor for the second is phosphoenol pyruvate **22**, and this yields cytidine-5'-triphosphate **23** or guanosine-5'-triphosphate **24**. Simultaneously, adenosine-5'-diphosphate **25** is regenerated to adenosine-5'-triphosphate **21**. Alternatively, the enzymes

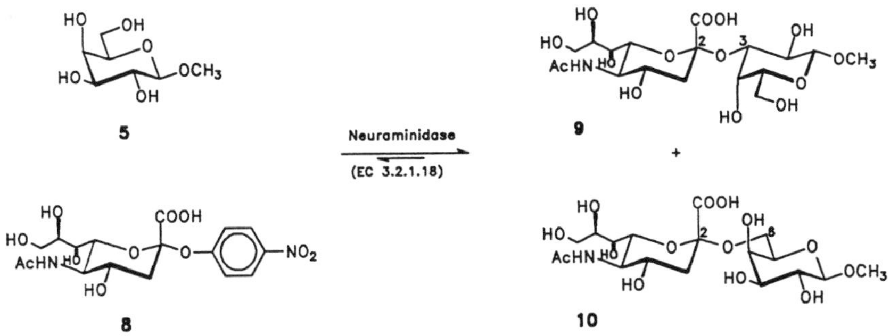

Scheme 6.   Disaccharide synthesis of Neu5Ac derivatives by catalysis with *vibrio cholerae* neuraminidase.

Table 2. Enzymatic glycosylation using vibrio cholerae neuraminidase[a]

Molar ratio pNPNeu5Ac(8)/ MeβGal(5)	Reaction time	Yield (%)	Ratio (2→3)-isomer/(2→6)-isomer (9) (10)	
A: free enzyme				
1 : 4	30 h	3	1 : 1.6	
1 : 15	25 h	5	1 : 9	
1 : 7	23 h	9	1 : 5.8	
B: immob. enzyme[b]				
1 : 7	42 h	16	1 : 2.5	

[a]Buffer: 0.05 M NaOAc, 0.001 M CaCl$_2$, pH 5.5
[b]Immobilisation on VA-Epoxy; activity yield: 75 %.

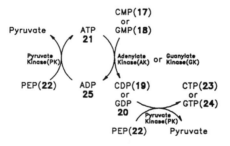

11: R¹=H, R²=OH,
 i=α-Glucosidase (EC 3.2.1.20)
15: R¹=OH, R²=H,
 i=α-Galactosidase (EC 3.2.1.22)

12: R¹=R¹'=H,
 R²=R²'=OH
16: R¹=R¹'=OH,
 R²=R²'=H

13

14

Scheme 7.   Enzymatic glycosylations using α-glucosidase and α-galactosidase.

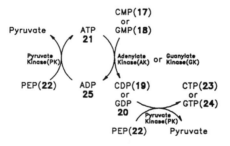

Scheme 8.   Combined regeneration of CTP or GTP.

were handled by various techniques: in the free form, using a dialysis bag for the reaction (membrane enclosed enzymic catalysis, MEEC, (*21*) or immobilized on the VA-epoxy support (*22*).   The latter allows a scale up to multigram amounts and yields between 42% to 79% of the described products (Table 3 and Table 4).   As yields determined by assays demonstrated, a considerable improvement may be expected when the workup and isolation of the product is modified.

In further improving these approaches towards Neu5Ac-linked oligosaccharides it is of interest to link the steps of the biological sequence to each other, rather than to disconnect them and isolate every intermediate.

This conception works out as shown in Scheme 10:   CTP **23** formed by the above described sequence is directly consumed by *N*-acetyl neuraminic acid **26** under the catalytic influence of cytidine-5'-monophosphosialate synthase (E.C. 2.7.7.43).   This enzyme is isolated from calf brain by ammonium sulfate precipitation (*23*) and subsequent affinity chromatography.   The stationary phase consists of CNBr-activated Sepharose 4B reacted with β-[3-(2-amino ethylthio)propyl]-*N*-acetyl neuraminic acid **27**, which is synthesized by radiating a mixture of the allyl glycoside and cysteamine to achieve radical C-S bond formation (*24*). The behavior of methyl β-*N*-acetyl-neuraminic acid as an inhibitor is in accordance with Zbiral's findings (*25*), where the methyl α-glycoside has been shown to compete with the native substrate for the enzyme, and thus **27** is recommended to be an ideally suited ligand (Scheme 9).   A typical analytical run is shown in Scheme 9.   Due to elution of the protein fraction by a salt gradient, the transfer to a preparative scale is rather difficult: denaturation occurs and thus a drop in activity down to 6% is observed.

The yields of the enzymatic reaction towards cytidine 5'-monophosphosialate **28** vary between 27 and 76% (*16.20.22.26*) (Scheme 10, upper half right) due to the reaction buffer chosen which is not yet fully understood.   In order to pull the reaction from the equilibrium state inorganic pyrophosphatase is added, which removes diphosphate from the mixture, and thus inhibits the reverse reaction to the starting materials **23** and **26**.

The coupling of both the synthesis of **23** (Scheme 8) and the synthesis of **28** (Scheme 10) is of major interest.   Not only would this be a complete reconstruction of the biological *in vivo* process, but it also is expected to give improved yields.   In Scheme 10 the combined reactions (*22*) are depicted: inexpensive enzymes such as e.g. pyruvate and adenylate kinase, respectively, are added in excess in order to overcome the phosphatase activity in the system, and to guarantee a sufficient CTP concentration.   Presently this approach gives the activated cytidine 5'-monophosphosialate **28** in merely 15.2%, which may be due to still remaining phosphatases.

The activated neuraminic acid **28** serves in transferase-catalyzed reactions as precursor for introduction of the terminal Neu5Ac residue in various oligosaccharides.   This approach is considerably more elegant than chemical synthesis, the drawback of which is a low-yielding glycosylation reaction with Neu5Ac glycosyl donors towards anomeric mixtures.   It was possible to link **28** by a sialyltransferase-catalyzed reaction (Scheme 11) to lactosamine (Galβ(1-4)GlcNAc, **29**) to yield the trisaccharide Neu5Acα(2-6)Galβ(1-4)GlcNAc **30** in 52% yield (*26*), and also reactions with the α(2-3) sialyltransferase could be carried out on an analytical scale (*27*).

Table 3. Different application of enzymes for regeneration of CTP

Type of Enzyme Application	Enzyme (U)	CMP (g)	Reaction Time (d)	Yield of CTP (%) (based on CMP) Isolation	Assay
Free	2500 AK 1000 PK	1.0	9	47	~80
MEEC Technique	5000 AK 2500 PK	1.0	10	51	~80
Immob. VA-Epoxy	200 AK 90 PK	0.5	5	42	~70

Table 4. Different application of enzymes for regeneration of GTP

Type of Enzyme Application	Enzyme (U)	GMP (g)	Reaction Time	Yield of GTP (%) (based on GMP) Isolation	Assay
Free	500 PK 5 GK	0.2	2 h	79	~80
MEEC Technique	500 PK 5 GK	0.3	5 d		36

N-Acetylneuraminic Acid (Neu5Ac, **26**)

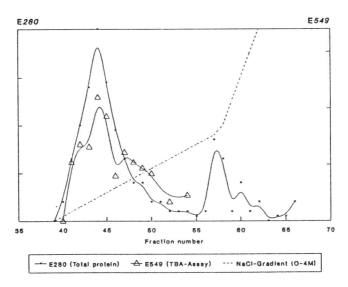

27

Scheme 9.   Purification of CMP-Neu5Ac-synthase by affinity chromatography.

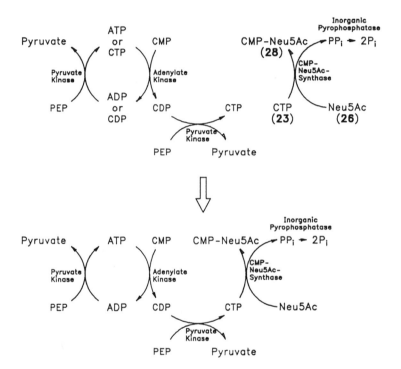

Scheme 10.   One-pot-synthesis of CMP-Neu5Ac with integrated CTP-regeneration.

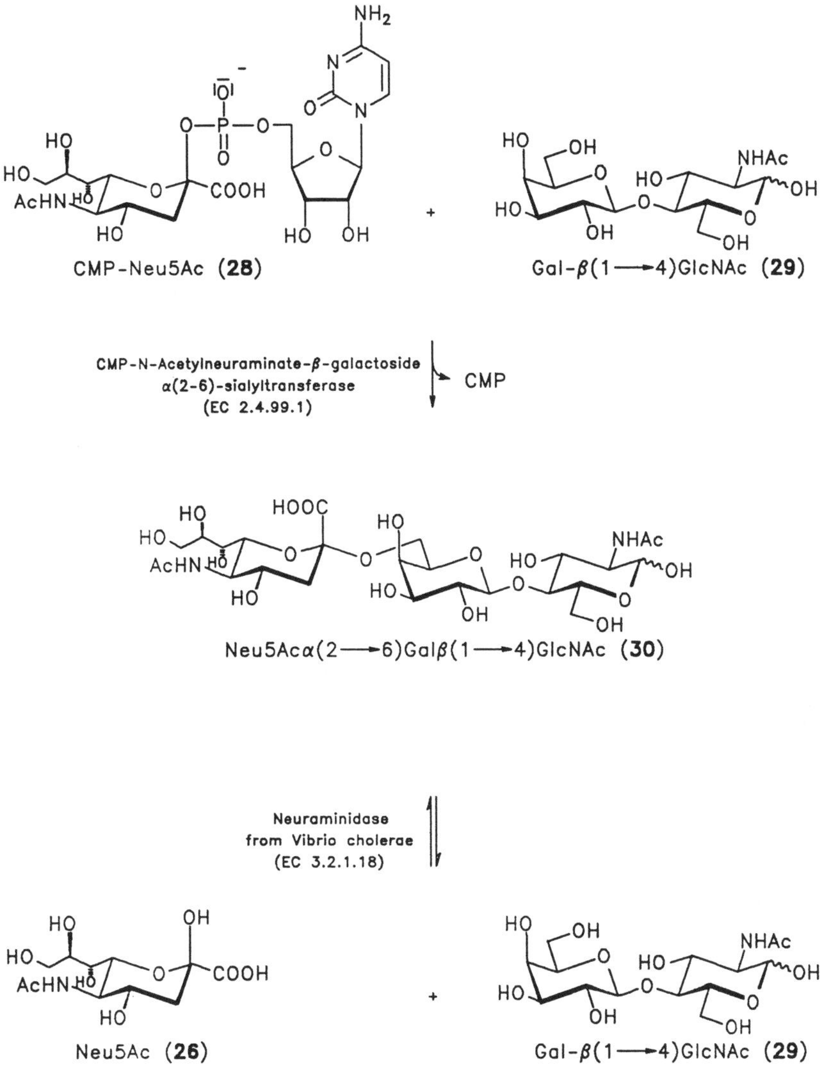

Scheme 11. Synthesis and cleavage of the trisaccharide Neu5Acα(2-6)Galβ(1-4)GlcNAc by transferases and hydrolases.

Fucosylation by a transferase enzyme is an analogous process to the transfer of neuraminic acid. L-Fucose **31** and ATP **21** are reacted in the presence of fucokinase (E.C. 2.7.1.52) to give fucose 1-phosphate **32**, and again the resulting ADP **25** is regenerated *in situ* by a phosphate transfer from phosphoenol pyruvate. The sugar phosphate **32** is converted to the corresponding nucleotide in a nucleotide transfer reaction using GTP **24**, catalyzed by GDP-fucopyrophosphorylase (E.C. 2.7.7.30). This gives guanosine 5'-diphosphofucose **33** in 32% yield, and again, pyrophosphate is cleaved to inorganic phosphate in the presence of inorganic pyrophosphatase (*28*) (Scheme 12).

By this process the activated fucose species is readily at hand for several fucosyltransferase-catalyzed reactions towards fucosylated oligosaccharides.

Galactosyltransferase reactions have been worked out to a much greater extent. As was demonstrated, this enzyme transfers galactosyl residues onto a variety of acceptors in approximately 30% yields (*16, 29-31*) (Scheme 13), and also the conception of cofactor regeneration (vide supra) was elaborated for such reactions. This is of major interest in the field of glycoprotein glycosylation, where attempts to synthesize larger units chemically are still extremely difficult and oligosaccharide structures of the complex or high mannose type have to be isolated either from milk, human faeces or swainsonine- and locoweed-intoxicated mammals, respectively. The first step in the direction of "man-designed" glycoproteins is demonstrated in the galactosylation of GlcNAc-asparagine methyl ester **34** the corresponding free acid derivative **35**, and chitobiosyl-asparagine methyl ester **3 6**, to give the corresponding di- and trisaccharide amino acids in approximately 30% yield (*31*).

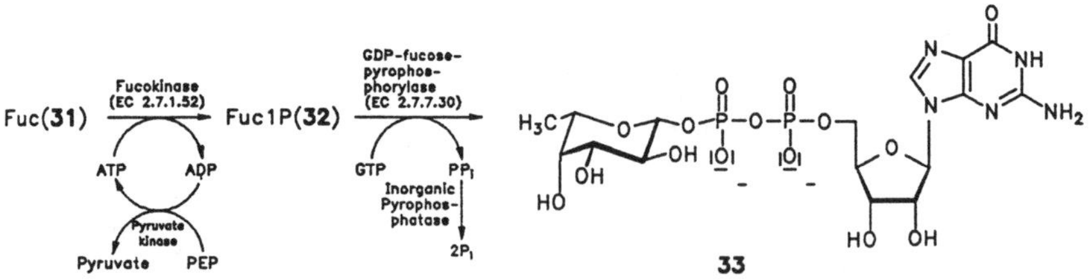

Scheme 12.   Synthesis of GDP-fucose.

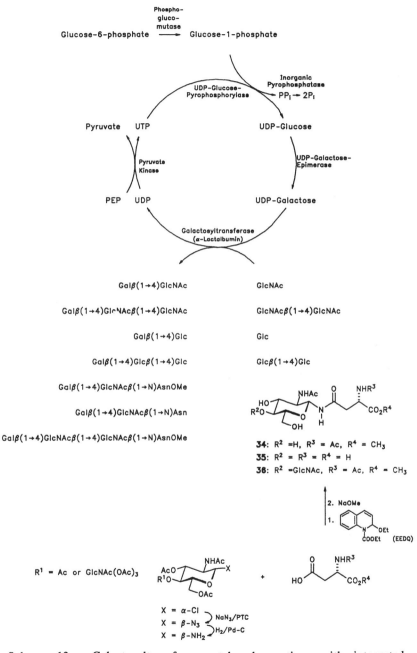

Scheme 13.   Galactosyltransferase-catalyzed reactions with integrated cofactor regeneration.

Acknowledgments

Studies of this group have enjoyed support from the *Deutsche Forschungsgemeinschaft*, the *Bundesministerium für Forschung und Technologie* and the *Fonds der Chemischen Industrie*, which is gratefully acknowledged.

Literature Cited

1. Rademacher, T. W.; Parekh, R. B.; Dwek, R. A. *Ann. Rev. Biochem.* **1988**, *57*, 785.
2. Hakomori, S. -I. *Ann. Rev. Biochem.* **1981**, *50*, 733.
3. Edelman, G. *Science* **1983**, *219*, 450.
4. Paulsen, H. *Angew, Chem. Int. Ed. Engl.* **1982**, *21*, 155.
5. Toone, E. J.; Simon, E. S.; Bednarski, M. D.; Whitesides, G. M. *Tetrahedron* **1989**, *45*, 365.
6. Nilsson, K. G. I. *Trends Biochem. Sci.* **1988**, *6*, 257.
7. Presper, K. A.; Concha-Slebe, I.; De, T.; Basu, S. *Carbohydr. Res.* **1986**, *155*, 73.
8. Alhadeff, J. A.; O'Brien, J. S. In *Practical Enzymology of the Shingolipidases*; Glew, R. H.; Peters, S. P., Eds.; Liss, New York, 1977, 247.
9. Van Hoof, F.; Hers, H. G. *Eur. J. Biochem.* **1968**, *7*, 34.
10. Svensson, S. C. T.; Thiem, J. *J. Carbohydr. Res.* **1990**, *200*, 391.
11. Giese, B.; Dupuis, J. *Angew. Chem. Int. Ed. Engl.* **1983**, *22*, 622.
12. Tsai, J. -H.; Berman, E. J. *Carbohydr. Res.* **1978**, *64*, 297.
13. Schauer, R. In *Sialic Acids Chemistry, Metabolism and Function*, Cell Biology Monographs, Springer-Verlag, Wien, 1982; Vol. 10.
14. Schauer, R. *Adv. Carbohydr. Chem. Biochem.* **1982**, *40*, 131.
15. Thiem, J.; Sauerbrei, B. submitted.
16. Thiem, J.; Treder, W.; Wiemann, T. *Formation of oligosaccharides by preparative enzymatic glycosylation*, Dechema Biotechnology-Conferences, 2, 189, VCH-Verlagsgesellschaft, Weinheim 1988.
17. Caputto, R.; Leloir, L. F.; Cardini, C. E.; Paladini, A. C. *J. Biol. Chem.* **1950**, *184*, 333.
18. Beyer, T. A.; Sadler, J. E.; Rearick, J. I.; Paulson, J. C.; Hill, R. L. *Adv. Enzymol.* **1981**, *52*, 24.
19. David, S.; Auge, C. *Pure Appl. Chem.* **1987**, *59*, 1501.
20. Simon, E. S., Bednarski, M. D.; Whitesides, G. M. *Tetrahedron Lett.* **1988**, *29*, 1123.
21. Bednarski, M. D.; Chenault, H. K., Simon, E. S.; Whitesides, G. M. *J. Am. Chem. Soc.* **1987**, *109*, 1283.
22. Thiem, J.; Stangier, P. *Liebigs Ann. Chem.*, in press.
23. Higa, H. H.; Paulson, J. C. *J. Biol. Chem.* **1985**, *260*, 8838.
24. Roy, R.; Laferriere, C. A. *Carbohydr. Res.* **1988**, *177*, C1.
25. Schmid, W.; Chrmstian, R.; Zbiral, E. *Tetrahedron Lett.* **1988**, *29*, 3643.
26. Thiem, J.; Treder, W. *Angew. Chem. Int. Ed. Engl.* **1986**, *25*, 1096.
27. Sabesan, S.; Paulson, J. C. *J. Am. Chem. Soc.* **1986**, *108*, 2068.
28. Thiem, J.; Stiller, R., unpublished.
29. Wong, C -H.; Haynie, S. H.; Whitesides, G. M. *J. Org. Chem.* **1986**, *108*, 158.
30. Auge, C.; David, S.; Mathieu, C.; Gautheron, C. *Tetrahedron Lett.* **1984**, *25*, 1467.
31. Thiem, J.; Wiemann, T. *Angew, Chem. Int. Ed. Engl.* **1990**, *29*, 80.

RECEIVED February 7, 1991

Chapter 6

Use of Glycosyltransferases for Drug Modification

Daniel Gygax[1], Mario Hammel[1], Robert Schneider[1],
Eric G. Berger[3], and Hiltrud Stierlin[2]

[1]Central Research Laboratories and [2]Research and Development
Department of Pharmaceuticals Division, Ciba-Geigy Ltd., 4002 Basel,
Switzerland
[3]Institute of Physiology, University of Zurich, Winterthurerstrasse 190,
8057 Zurich, Switzerland

Glycosyltransferases are enzymes which form glycosidic linkages
regio- and stereospecifically by transferring the carbohydrate residues
of activated monosaccharides to suitable acceptors. In this paper we
describe the use of glycosyltransferases for the in vitro modification of
drugs. Glycosyltransferases were used for the modification of
glycoprotein glycans expressed in yeast. A continuous synthesis of β-
D-glucuronides catalyzed by glucuronyltransferases was carried out in
a membrane reactor. UDP-glucuronic acid was regenerated in situ by
using a multi-enzyme system. Modification of glycan chains was
performed by combined use of glycosidases and glycosyltransferases.
Sialylated glycoproteins showed extended plasma residence time
compared to galactosylated glycoproteins.

The selective synthesis of β-glucuronides with glucuronyltransferases
provides a method to construct β-D-glucuronides of xenobiotics such as
drugs, pesticides and endogenous steroid hormones which are widespread
biotransformation products. These conjugates are needed for analytical,
toxicological and pharmacological investigations. To perform these studies,
amounts between 50 to 100 mg of radioactively labeled or unlabeled
glucuronide are required.
    The enzyme-catalyzed synthesis of β-D-glucuronides is an
alternative to the chemical synthesis or the isolation of these conjugates
from biological fluids. The UDP-glucuronyltransferases (EC 2.4.1.17) are a
family of enzymes located predominantly in endomembranes of the
hepatocyte (1). They catalyze the transfer of glucuronic acid from uridine
5'-diphospho-α-D-glucuronic acid to a suitable aglycon with inversion of
the configuration at the anomeric center of glucuronic acid (Figure 1) (1).
The reaction is stereoselective; only the β-glucuronide is formed. None of
the functional groups of glucuronic acid have to be protected. The scope of
the reaction is broad: glucuronyltransferase was found to conjugate
aglycons containing phenols, amines, alcohols, thiols, carbamates and
carboxylic acids (2).

0097–6156/91/0466–0079$06.00/0

Synthesis   of   Glucuronides

Glucuronides have been synthesized batch-wise or in a hollow fiber system using microsomal or soluble enzyme preparations (3-5). Furthermore, they have been prepared with enzymes immobilized to polymeric supports (6). Here we describe the continuous synthesis of glucuronide conjugates in a 10-mL membrane reactor (7).

Experimental   Details.   Figure 1 depicts the general reaction scheme. Fresh liver from guinea pig was homogenized and centrifuged. A 1-mL portion of the supernatant was injected directly into the reactor (Figure 2). There it was retained by a semi-permeable membrane with a molecular weight limit of 10 kD. The substrate solution, containing 5 mM of the ^{14}C-labeled aglycon (1 or 2), 20 mM of UDP-glucuronic acid, 10 mM of MgCl$_2$ and 50 mM of HEPES buffer (pH 7.4) was pumped through the reactor at a flow rate of 6 mL/h. The glucuronide production profile of **1** reached a plateau corresponding to 95% conversion within 8 hours. This high level of conversion was maintained throughout the whole reaction period of 20 hours. Compound **2** was a poorer substrate that **2**, and, consequently, resulted in only about 50% conversion over the reaction period of 20 h. Analysis by ^{1}H-NMR spectroscopy revealed that the enzyme catalyzes the conjugation of **2** stereo- and regioselectively by differentiating between the aryl- and the alkylhydroxy groups. The formation of **1a** and **2a** were followed by TLC and quantified by radioactivity scanning. After purification using reversed-phase HPLC, 37 mg (50%) of **1a** and 96 mg (37%) of **2a** were isolated. Structural and stereochemical assignments were based on analysis of the ^{1}H-NMR and MS spectra.

The glucuronyltransferase-catalyzed reaction is species dependent (Table 1). We used a crude liver homogenate from various animal species as the source of the enzymes.

Regeneration   of   UDP-Glucuronic   Acid.   The   substrate   of enzymatic glucuronidation, UDP-glucuronic acid, is very costly and unstable. Therefore, we established a multi-enzyme system to regenerate UDP-glucuronic acid in situ from   uridine 5'-diphosphate (UDP) and glucose 1-phosphate (Figure 3). A similar type of multi-catalyst system with immobilized enzymes was used for the synthesis of oligosaccharides (8-9). Instead of using individually immobilized enzymes, we used a crude liver homogenate from guinea pig containing all enzymes involved in the multi-catalyst system.

The glucuronidation is started with 0.05 equivalents of UDP-glucose compared to the aglycon. UDP-glucose is oxidized by the NAD$^+$-dependent UDP-glucose dehydrogenase to UDP-glucuronic acid (Step II). Two products are generated during UDP-glucuronyltransferase-catalyzed transfer of glucuronic acid to the aglycon (Step I), namely the glucuronide and UDP. The phosphorylation of UDP to UTP is catalyzed by pyruvate kinase (Step IV) using phosphoenolpyruvate (PEP) as phosphoryl-group donor. Finally, the cycle is closed by the UDP-glucose pyrophosphorylase-catalyzed transfer of UTP to glucose 1-phosphate (Step III).

Scale-up   of   Procedure.   The multi-catalyst synthesis of **1a** was established and optimized on an analytical scale. The formation of **1a** was followed by TLC and quantified by radioactivity scanning. Optimal conditions were: 0.5 mM of **1**, 0.025 mM of UDP-glucose, 3.3 mM of glucose 1-phosphate, 2 mM of PEP, 2 mM of NAD$^+$ and 20 mg/mL of liver homogenate in 0.025 M of HEPES buffer (pH 8.0). In preparative scale

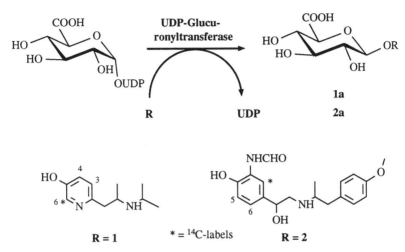

Figure 1. Enzymatic synthesis of β-D-glucuronides with crude liver homogenate from guinea pig as source of the glucuronyltransferase. **1**: CGS 5649 B. **2**: CGP 25 827 A.

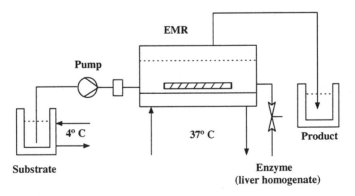

Figure 2. Schematic representation of the enzyme membrane reactor (EMR) system used for enzymatic synthesis of β-D-glucuronides.

Table 1.   Species dependency of enzymatic glucuronidation

Order	Species	Conversion (%)	
		CGS 5649	CGP 361
	Man	<5	5
Non-human primates	Baboon	10	35
	Marmoset	0	15
Lagomorphae	Rabbit	100	85
Rodents	Rat	70	20
	Mouse	95	15
	Guinea-pig	100	25
	Syr. Hamster	90	50
	Chin. Hamster	100	45
Carnivores	Dog	5	15
Odd-toed ungulates	Horse	100	30
Even-toed ungulates	Calf	50	20
	Goat	80	20
	Sheep	100	35
	Pig	100	25
Gallinaceae	Hen	40	15
Salmonidae	Troute	35	20
Molluscs	Arion hort.	0	15

CGS 5649   CGP 361

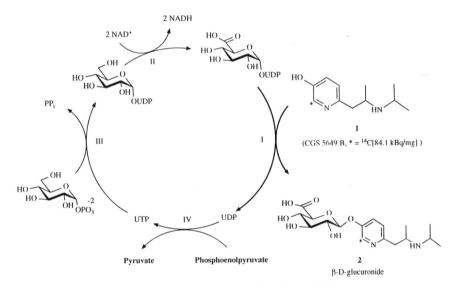

Figure 3.   Enzymatic   glucuronidation   of   CGS   5649   B   with   *in   situ* regeneration   of   UDP   and   glucose   1-phosphate.

experiment, the conjugation of glucuronic acid to CGS 5649 B yielded 45 mg (65%) of **1a**. Structural and stereochemical assignments of **1a** were based on analysis of ^{1}H-NMR and MS spectra.

In Vitro Modification of Glycoprotein Glycans

Recombinant proteins which are expressed in yeast may not be suitable as drugs for in vivo application since the *N*-linked glycan is of the high- or polymannose type, whereas the mammalian glycan is of the complex type (*10*). In the late sixties, Ashwell and Morell found that the sugar composition of a glycoprotein determines its pharmacokinetics (*11*). They showed that serum glycoprotein glycans with a terminal galactose bind to the hepatic asialoglycoprotein receptor and are internalized into the cell. The presence of *N*-acetylneuraminic acid as terminal sugar prevents binding to the receptor, and subsequent endocytosis; consequently, plasma half-life is extended. In the light of the existence of other sugar-binding receptors, e.g. mannosyl receptor of the reticuloendothelial system, methods should be designed to properly glycosylate drugs produced in heterologous systems.

   Our strategy is to modify glycoprotein glycans expressed in yeast by in vitro use of both glycosidases and glycosyltransferases (*12-14*). The aim is to synthesize a mimic of the outer chain of complex *N*-glycans. This can be achieved by deglycosylating the high-mannose type and by adding galactose and *N*-acetylneuraminic acid to the truncated yeast glycoprotein (Figure 4). This procedure ensures glycosylation at the proper place of the polypeptide without its denaturation.

   In our study we used four different glycoprotein acceptors which all have different numbers of carbohydrate acceptor sites per molecule of protein: ovalbumin (0.9 mol of *N*-acetylglucosamine/mol protein), a chemically modified BSA (35 mol of *N*-acetyl-glucosamine/mol protein), yeast invertase (9 mol of polymannan/mol protein) and a recombinant tPA B-chain variant expressed in yeast (1 mol of polymannan/mol protein).

   Enzyme-catalyzed Steps. Deglycosylation was the first step of the modification. Deglycosylation of the polymannan invertase is achieved by endoglycosidase H (*15*). This enzyme hydrolyzes the β-1,4 linkage between the two *N*-acetylglucosamine residues of the core glycan. Thirty mU of endoglycosidase H catalyzed the hydrolysis of about 40 mg of yeast invertase at 25°C within 24 h. The progress of the reaction was monitored by SDS-PAGE. The product is a truncated glycoprotein with a molecular weight of 62 kD, containing 7 out of 9 possible *N*-acetylglucosamine residues linked to the asparagine residues of invertase.

   In a second step, β-1,4-galactosyltransferase (70 mU) catalyzed the transfer of galactose from the ^{14}C-labeled UDP-activated galactose to the acetylglucosamine residue of the truncated invertase (20 mg) (*12*). The reaction was performed at room temperature and resulted after 50 h in a galactosylation efficiency of about 70%.

   In the third step, the galactosylated invertase (20 mg) was condensed with ^{14}C-labeled *N*-acetylneuraminic acid in a reaction catalyzed by sialyltransferase (40 mU) at room temperature (*12, 16-17*). The sialyltransferase catalyzes the formation of an α-2,6 linkage. Approximately 70-80% of the available acceptor sites of invertase were sialylated. This modification could be performed very efficiently with soluble enzymes. However, for larger scale production, immobilized

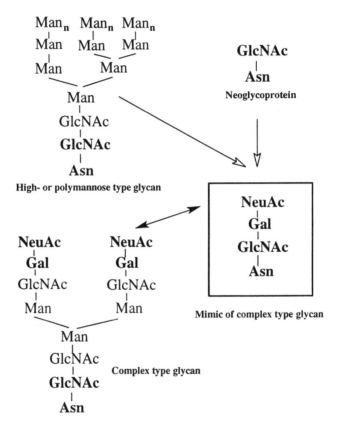

Figure 4.   Schematic representation of the *in vitro* synthesis of a trisaccharide outer chain of complex *N*-glycans.

catalysts   would   be   preferable.   To   study   this   approach,
galactosyltransferase   was   immobilized   on   carriers   and   galactosylation   was
continuously carried out in a reactor.
Immobilization   of   Enzymes.   Galactosyltransferase   was
immobilized   to   CNBr-   or   tresylchloride-activated   Sepharose(*18*),   or   to
Concanavalin   A-Sepharose   by   virtue   of   its   *N*-glycans   (*19-20*).
Galactosyltransferase   was   also   immobilized   to   monoclonal   anti-
galactosyltransferase   antibodies (*21*)   which   were   bound   via   their   Fc-
fragment to Protein G-Sepharose (*22*).
The   various   immobilization   methods   showed   different
characteristics.   With the covalent methods, up to 72% of the enzyme could
be bound to the carrier, but specific activity dropped to less than 10%.   In
contrast,   noncovalent   immobilization   yielded   only   about   50%
immobilization   efficiency,   but   21%   (Con A)   and   25%   (monoclonal   anti-
galactosyltransferase antibody) of specific activity could be recovered.
The "slurry" reactor, which was introduced by Bossow and Wandrey
for   enzymatic   formation   of   C-C   bonds   (*23*),   is   basically   a   stirred
ultrafiltration   cell   (10   mL)   containing   a   membrane   with   a   defined
molecular   weight   limit   of   100   or   300   kD,   which   serves   as   barrier.   The
components   of   the   galactosylation   mixture   are   continuously   pumped
through   the   reactor   and   pass   through   this   barrier,   whereas   the
immobilized catalyst is retained in the reactor.   The immobilized enzyme is
held in suspension by continuous stirring.
Galactosylation   Reactions.   For   continuous   galactosylation   of
ovalbumin   (10   mg/mL),   74   mU   of   the   Con   A-immobilized
galactosyltransferase were used.   The mean residence time of the substrates
in the reactor was about 2.7 h.   The reaction was run for 27 h and resulted
in a maximum galactosylation efficiency of 55% (Figure 5).
BSA-GlcNAc is bovine serum albumin modified with spacer arms
which   carry   terminal   GlcNAc   residues.   36   mU   of   the   Con   A-immobilized
galactosyltransferase were used to continuously galactosylate BSA-GlcNAc
during   23   h   with   a   mean   residence   time   for   the   substrates   of   5.3   h.   The
galactosylation efficiency reached a maximum of 25% after about 11 h and
decreased slightly during the steady-state period (Figure 5).
The galactosylation of endo H-treated invertase (1 mg/mL) was run
in a reactor with a volume of 8 mL containing 80 mU galactosyltransferase
immobilized   to   tresylchloride-activated   Sepharose.   The   mean   residence
time of the substrates in the reactor was 10.6 h.   The reaction was first run
for 22 h and proceeded with an efficiency of 35%.   In a second run, the
galactosylation efficiency was improved to about 75% (Figure 6).
The   stability   of   the   immobilized   galactosyltransferase   was
considerably better than that of the soluble enzyme.   At room temperature
only 55% of initial immobilized activity was lost within 65 hours compared
to a loss of 95% of the activity of the soluble enzyme.

**Synthesis   of   Labeled   Neoglycoproteins   to   Determine   Plasma
Elimination   Profiles**

Using soluble β-1,4-galactosyltransferase and α-2,6-sialyltransferase, we
synthesized glycan "mimics" of BSA-GlcNAc and yeast tPA.   Each glycan
mimic of both neoglycoproteins was terminated with either ^{14}C-labeled
galactose or ^{14}C-labeled *N*-actelyneuraminic acid.   These radioactively-
labeled neoglycoproteins were injected into rats in order to determine
their plasma elimination profiles.

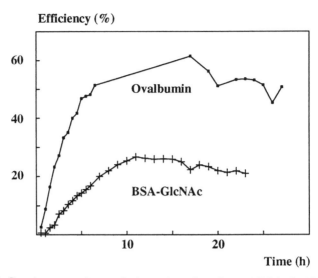

Figure 5. Continuous galactosylation of ovalbumin and BSA-GlcNAc in the reactor. The galactosyltransferase was immobilized to Concanavalin A-Sepharose. Efficiency is defined as concentration of galactosylated acceptor in percent of total acceptor concentration. (Reprinted with permission from ref. 24. Copyright 1990 Glycoconjugate Journal.)

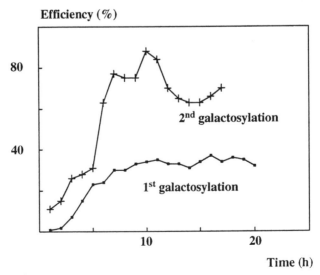

Figure 6. Continuous galactosylation of endo H-treated invertase in a reactor containing the galactosyltransferase immobilized to tresylchloride-activated Sepharose. The product of the first run was galactosylated a second time under similar conditions. (Reprinted with permission from ref. 24. Copyright 1990 Glycoconjugate Journal.)

As expected, the galactosylated neoglycoproteins of BSA and tPA were cleared much more rapidly from the plasma than the sialylated neoglycoproteins (Figure 7).   In addition to the plasma elimination profile, we also wanted to know in which tissues the neoglycoproteins distributed. After 20 minutes most of the radioactivity was found in the liver.   More of galactosylated than sialylated neoglycoproteins were found in this organ (Figure 8).

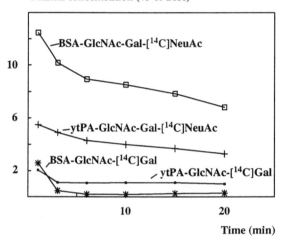

Figure 7.   Plasma elimination profile of yeast tPA B-chain variant and BSA-neoglycoprotein.   The enzymatically modified neoglycoproteins (about 1 x 10^6 dpm) were injected into rats.

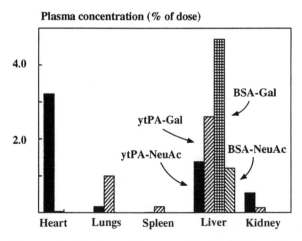

Figure 8.   Organ distribution of yeast tPA B-chain variant and BSA-neoglycoprotein.

Conclusion

In conclusion, reactions catalyzed by glucuronyltransferases are stereoselective but accept a wide variety of substrates. Preparative-scale synthesis can be performed in a membrane reactor, and costly UDP-glucuronic acid can be regenerated *in situ*. Modification of glycan chains is feasible by combined use of glycosidases and glycosyltransferases. Galactosylation can be carried out continuously in a slurry reactor. Sialylated glycoproteins show extended plasma elimination half-lives compared to galactosylated glycoproteins.

Literature   Cited

1. Zakim, D.; Hochman, Y.; Vessey, D. A. In *Biochemical Pharmacology and Toxicology*; Zakim, D.; Vessey, D. A., Eds.; John Wiley and Sons: New York, 1985; Vol. 1, p 161.
2. Burchell, B.; Coughtrie, M. W. H. *Pharma. Ther.* 1989, *43*, 261-289.
3. Johnson, D. B.; Swanson, M. J.; Barker, C. W.; Fanska, C. B.; Murrill, E. E. *Preparative Biochemistry* 1979, *9*, 391-406.
4. Guenthner, T. M.; Blair, N. P. *Pharmacology* 1988, *37*, 341-348.
5. Tegtmeier, F.; Belsner, K.; Brunner, G. *Bioprocess Engineering* 1988, *3*, 43-47.
6. Dulik, D. M.; Fenselau, C. *FASEB J.* 1988, *2*, 2235-2240.
7. Gygax, D.; Nachtegaal, H.; Ghisalba, O.; Lattmann, R.; Schär, H. -P.; Wandrey, C.; Streiff, M. B. *Appl. Microbiol. Biotechnol.* 1990, *32*, 621-626.
8. Wong, C. -H.; Haynie, S. L.; Whitesides, G. M. *J. Org. Chem.* 1982, *47*, 5418-5420.
9. Augé, C.; David, S.; Mathieu, C.; Gautheron, C. *Tetrahedron Letters* 1984, *25*, 1467-1470.
10. Kukuruzinska, M. A.; Bergh, M. L. E.; Jackson, B. J. *Annu. Rev. Biochem.* 1987, *56*, 915-944.
11. Ashwell, G.; Morell, A. G. In *Adv. Enzymology* 1974, *41*, 99-128.
12. Berger, E. G.; Greber, U. F.; Mosbach, K. *FEBS Letters* 1986, *203*, 64-68.
13. Toone, E. J.; Simon, E. S.; Bednarski, M. D.; Whitesides, G. M. *Tetrahedron* 1989, *45*, 5365-5422.
14. Thotakura, N. R.; Bahl, O. P. *Methods Enzymology* 1987, *138*, 350-359.
15. Trimble, R. B.; Maley, F. *J. Biol. Chem.* 1977, *252*, 4409-4412.
16. Higa, H. H.; Paulson, J. C. *J. Biol. Chem.* 1985, *260*, 8838-8849.
17. Thiem, J.; Treder, W. *Angew. Chemie* 1986, *12*, 1100-1101.
18. Nilson, K.; Mosbach, K. *Methods Enzymology* 1987, 135, 65-79.
19. Endo, T.; Amano, J.; Berger, E. G.; Kobata, A. *Carbohydr. Res.* 1986, *150*, 6882-6887.
20. Berger, E. G. In *Colloquium Protides of the Biological Fluids*; Pergamon Press: 1986; p 799.
21. Berger, E. G.; Aegerter, E.; Mandel, T.; Hauri, H. -P. *Carbohydrate Res.* 1986, *149*, 23-33.
22. Solomon, B.; Koppel, R.; Pines, G.; Katchalski-Katzir, E. *Biotechnol. Bioeng.* 1986, *23*, 1213-1221.
23. Bossow, B.; Berke, W.; Wandrey, C. In *VDI-Gesellschaft für Verfahrenstechnik*; 1988.
24. Schneider, R.; Hammel, M.; Berger, E. G.; Ghisalba, O.; Nueesch, J.; Gygax D. *Glycoconjugate J.* 1990, *17*, 589–600.

RECEIVED February 1, 1991

Chapter 7

Some Aspects of Enzymic Synthesis of Oligosaccharides Employing Acceptor Saccharides Attached to Polymer Carriers

Uri Zehavi[1] and Joachim Thiem[2]

[1]Faculty of Agriculture, The Hebrew University of Jerusalem, Rehovot 76100, Israel
[2]Institut für Organische Chemie, Universität Hamburg, D-2000 Hamburg 13, Germany

This chapter discusses recent examples of enzymic oligosaccharide synthesis on polymer supports. Examples include cases where water-insoluble acceptor-polymers of improved accessibility are used with soluble enzymes, and cases where water-soluble acceptor-polymers are used with either insoluble or soluble enzymes. The methodology is useful for preparative-scale reactions and for the study of enzymic properties.

Oligosaccharides are of prime biological importance. Although advances in their chemical synthesis have been remarkable in recent years (1), many structures are still extremely difficult to prepare. Although less developed in the preparative sense, numerous useful applications of enzymic syntheses of oligosaccharides were discussed in a recent symposium (2). The enzymic synthesis of oligosaccharides on polymer supports (3) is an excellent general approach and exploits the specificity of glycosyltransferases or transglycosylation reactions in determining the nature of the newly-formed glycosidic bond. Complex protecting group chemistry is avoided, and at the end of the reaction the products can be released from the polymer and purified. The present article includes some recent data that illustrates the relevance of this methodology in the preparation of oligosaccharides and in the study of the properties of enzymes.

Insoluble Acceptor-polymers and Soluble Enzymes

In our prior experience, employing aminoethyl-substituted poly(acrylamide) gel beads as acceptors in the galactosyltransferase reaction, the transfer yields were lower than 1% compared to a 30% yield in the case of a soluble acceptor-polymer (3). Longer spacer arms were tried, with particular attention to the swelling of the resulting polymers, and the

0097–6156/91/0466–0090$06.00/0

D-galactopyranosyl transfer yield with the same enzyme went up to 8.7% in the case of aminohexyl-substituted poly (acrylamide) gel beads carrying β-D-glucopyranosyl residues, **1** (Figure 1) (*4*). The ease with which products are isolated by this polymer technique is noteworthy. It permits efficient comparison of acceptor specificity by employing a variety of acceptors, and is capable of demonstrating *de novo* synthesis even in cases of extremely low condensation yields. This has been utilized by us in the case of glycogen synthase (*5*) and was extended now to acceptor polymers carrying both α1-4 and α1-6-D-glycopyranosyl substituents (*6*). Thus, starting from the same aminohexyl-substituted light-sensitive polymer, polymers carrying maltose **2**, maltotriose **3**, a glycogen-branch point trisaccharide **4** and panose **5** were synthesized and served in a comparative study as acceptors in the glycogen synthase reaction. Highest transfer was observed with the *maltrio* polymer, acceptor efficiencies being 3>2>>5>4 (Table I). While extending the acceptor with α1-4-D-glucopyranosyl residues improved the rate of transfer, the inclusion of α1-6-D-glucopyranosyl residues decreased the rate of transfer. The acceptor efficiency seems to improve with lengthening the maltodextrin units when using the bulkier glycogen synthase. The extent of polymerization was assayed by monitoring the incorporation of [U-^{14}C] glucose from UDP-[U-^{14}C] glucose into compounds **2-5**. Subsequently, product polymers obtained following incorporation of [U-^{14}C] glucose into compounds **3, 4**, and **5** released upon irradiation 16, 92 and 12% of the radioactivity (Table I, Experiment 2), respectively. The unreleased products appear to be of higher molecular weight than the starting materials (*5*). In general, insoluble acceptor-polymers are very useful for the determination of the extent of transfer and the study of primer requirements. Experimental details for experiment 1 include small modifications of those of Zehavi and Herchman (*5*). In experiment 2, incubation was carried out as in experiment 1 but in ultrafiltration cells equipped with Diaflo PM-10 membranes (Amicon, Lexington MA 02173). Following 6 and 12h of incubation most of the solution was removed by filtration and the remainder washed with buffer (Tris-HCl, 50mM, pH 7.8). Incubation mixture and glycogen synthase were added and the incubation was continued for a total of 18 h. Indicated are quantities of glucose added to the corresponding polymer (μmol/g) upon transfer. Another example concerned with the study of enzyme specificity is that of cyclodextrin α1-4 glucosyltransferase (CGT). The known reactions of CGT are disproportionation, cyclization and coupling (Figure 2).

Presently, by using α-cyclodextrin as the donor and polymer **1** (Figure 1) as the acceptor, we were able to select for and demonstrate disproportionation and coupling (increased formation of maltoheptaose following release from the product polymer, **6**) even though under the conditions used the coupling yield was very low. Cyclization products (cyclodextrins), if formed, were not attached to the polymer (Zehavi, Thiem, and Herchman, to be published, Table II). The following paragraph provides a representative experimental procedure.

Polymer **1** (50 mg, 210 μmol of glucose/g) and α-cyclodextrin (40 mg) were preincubated in 0.05 M sodium acetate buffer (pH 6.0, 1.5 ml) at 45°. Following complete swelling, cyclodextrin α1-4 glucosyltransferase (1 mg, 18 units, from *Klebsiella pneumoniae)* dissolved in the same buffer (0.5 ml) was added and the incubation was continued for 10 min. The product polymer (**6**) was centrifuged and washed extensively with water until no

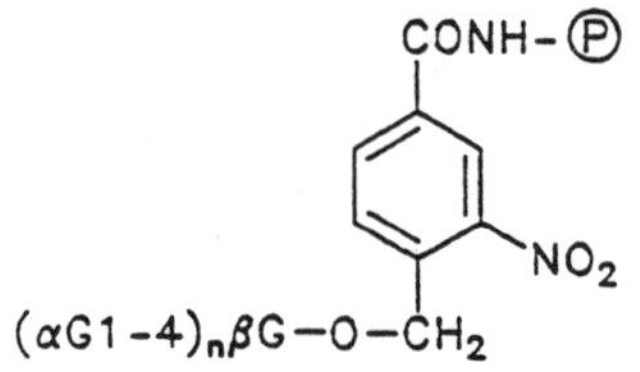

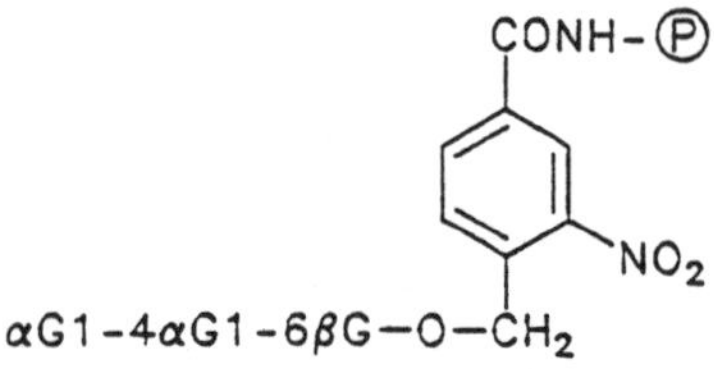

1 n=0
2 n=1
3 n=2
6 n 0-8

4

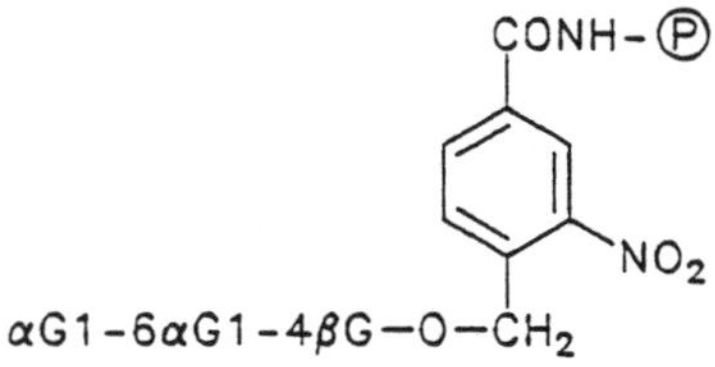

5

G = D-glucopyranosyl
Ⓟ = aminohexyl-substituted poly(acrylamide) beads

Figure 1. Insoluble acceptor polymers in the galactosyltransferase (*1*), cyclodextrin α1-4 glucosyltransferase (*1*) and in the glycogen synthase reactions (*2-5*).

Table I. Glycogen Synthase-Catalyzed Incorporation of D-Glucose into Acceptor Polymers

	Experiment 1 µmol/g	% transfer	Experiment 2 µmol/g	% transfer
2	1.58	0.63		
3	0.58	0.87	5.20	7.76
4	0.11	0.16	0.17	0.24
5	0.28	0.20	0.96	0.68

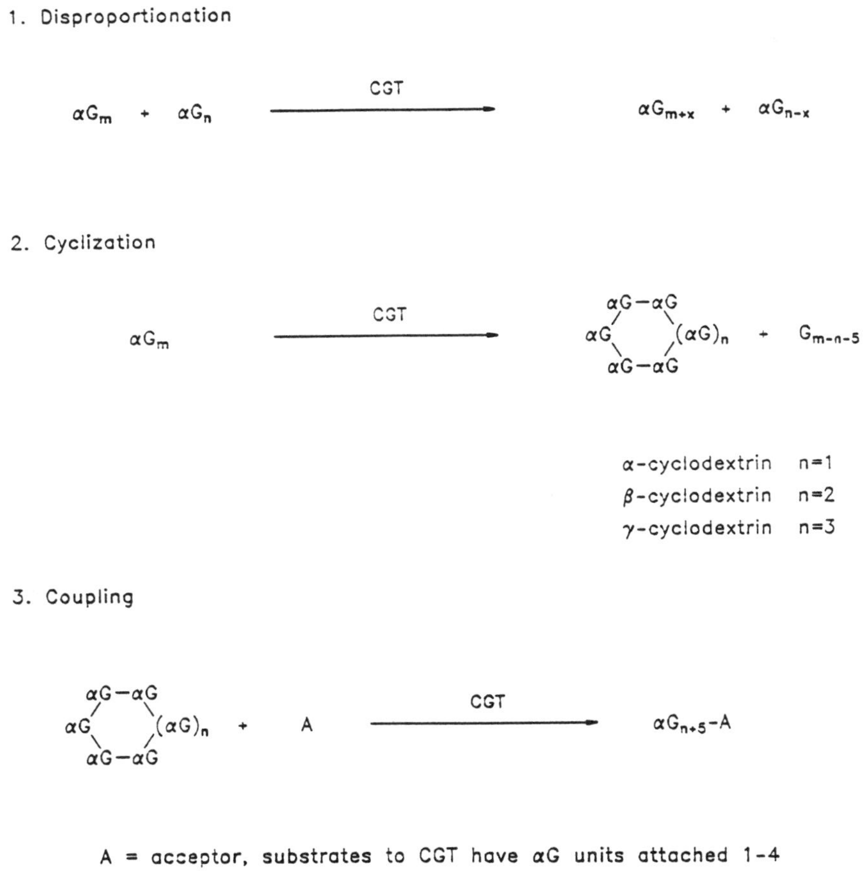

1. Disproportionation

$$\alpha G_m \;+\; \alpha G_n \xrightarrow{\text{CGT}} \alpha G_{m+x} \;+\; \alpha G_{n-x}$$

2. Cyclization

$$\alpha G_m \xrightarrow{\text{CGT}} \begin{array}{c} \alpha G - \alpha G \\ \alpha G \qquad (\alpha G)_n \\ \alpha G - \alpha G \end{array} \;+\; G_{m-n-5}$$

$$\begin{array}{ll} \alpha\text{-cyclodextrin} & n=1 \\ \beta\text{-cyclodextrin} & n=2 \\ \gamma\text{-cyclodextrin} & n=3 \end{array}$$

3. Coupling

$$\begin{array}{c} \alpha G - \alpha G \\ \alpha G \qquad (\alpha G)_n \\ \alpha G - \alpha G \end{array} \;+\; A \xrightarrow{\text{CGT}} \alpha G_{n+5} - A$$

A = acceptor, substrates to CGT have αG units attached 1-4

Figure 2.   Reactions catalyzed by cyclodextrin α1-4   glucosyltransferase.

Table II. Photochemical Release of Oligosaccharides from
Polymer 6

Oligosaccharide	μmol/g	Yield (%)
Glucose	185	88
Maltose	1.53	0.73
Maltotriose	1.68	0.80
Maltotetraose	1.96	0.93
Maltopentaose	0.98	0.47
Maltohexaose	0.11	0.05
Maltoheptaose	0.21	0.10
Maltooctaose	<0.02	<0.01

oligosaccharides could be detected in the supernatant). It was then suspended in water (10 ml) and irradiated for 17 h at room temperature in a Reyonet RPR-100 apparatus with RPR 3500 Å lamps. The released oligosaccharides were determined (μmol/g of polymer **6** following centrifugation and concentration of the supernatant) by HPLC (7, 8).

Water-soluble Acceptor-Polymers and Insoluble Enzymes

Based on prior experiences where immobilized cyclodextrin $\alpha 1-4$ glucosyltransferase (iCGT) was prepared and used with α-D-glucopyranosyl fluoride (GF) as the donor (7), we have demonstrated the iCGT catalyzed incorporation of α-D-glucopyranosyl units (G) to polymer-bound maltodextrins (PBM, the polymer being water-soluble substituted poly(acrylamide), **8**). It has been predicted that using a soluble polymer bearing a saccharide acceptor as a handle has the advantage of selecting for disproportionation products. Thus, with a high excess of activated monosaccharide donor (GF) and a soluble polymer it will be possible to isolate polymer-bound products in the presence of nonbound low molecular weight components, e.g., reactants, products, and byproducts. The increase in the saccharide content in polymers **9** and **10** was of 7.8 and 5 fold, and the relase of saccharides was performed photochemically (Figure 3).

HPLC comparison of saccharides **11** and **12** suggest that both are apparently equilibrium mixtures composed of similar proportions of maltooligosaccharides (dp ≥ 2) while no significant hydrolysis of the *malto* substituent in polymer **8** to a *gluco* substituent is observed. Moreover, as predicted, saccharides **11** and **12** were disproportionation products; cyclization products could not be attached to polymers **9** and **10** and were removed by dialysis.

A Soluble Acceptor-Polymer and a Soluble Enzyme

By glycosylation and further condensation reactions along previously worked out conceptions, the attachment of *N*-acetyl glucosamine to aminoethylated-substituted poly(acrylamide) led to the polymer-bound derivative **14**. As demonstrated in other recent experiments (9), it was advantageous to perform the enzymic galactosylation by use of *in situ* generated UDP-galactose. This in turn was obtained from much cheaper UDP-glucose in the presence of UDP-galactose epimerase. Further, treatment of the reaction mixture with D-galactosyltransferase gave smoothly the galactosylated compound **15**. Finally, the polymer-bound *N*-acetyl lactosamine was released from the light-sensitive support by irradiation to give *N*-acetyl lactosamine (**16**) in approximately 2% overall yield with respect to enzymic glycosylation, cleavage by irradiation and isolation by ultrafiltration (Thiem, Wiemann, and Zehavi, unpublished results) (Figure 4).

Glycosphingolipids, being important constituents of biological membranes, are obvious targets for synthesis on polymer supports. On the assumption that glycosyl sphingosine derivatives, which are attached through a temporary protecting group of the 2-amino function to a carrier polymer, may serve as acceptors for glycosyltransferases involved in the biosynthesis of glycosphingolipids, 4-carboxymethyl-2-nitrobenzyl-

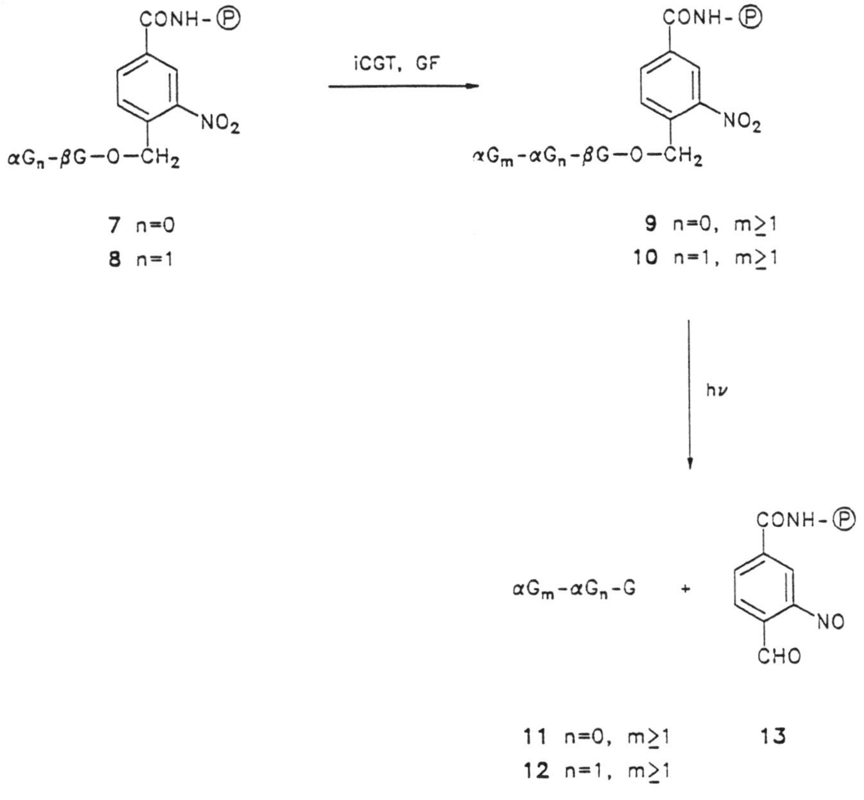

Ⓟ = water-soluble substituted poly(acrylamide)

Figure 3. Immobilized α1-4 glucosyltransferase reaction with soluble polymeric acceptors followed by photochemical release of saccharides from the polymers. M indicates glycosyl residues added by transfer.

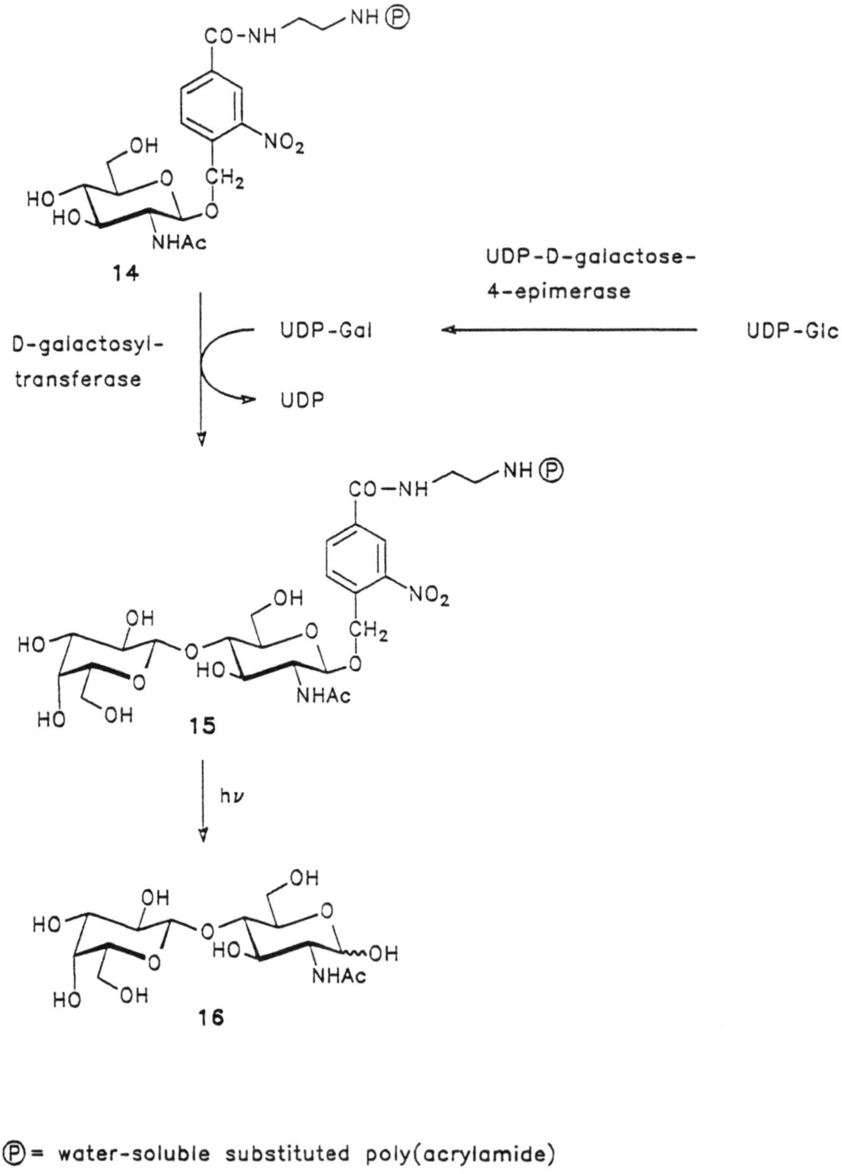

$\textcircled{P}$ = water-soluble substituted poly(acrylamide)

Figure 4.   Galactosyltransferase reaction with a solution polymeric acceptor followed by photochemical release of the product from the polymer.

oxycarbonyl was studied as a bifunctional protecting group which aims to serve this purpose (*10*).

Starting from synthetic (2*S*, 3*R*, 4*E*)-2-amino-1-β-D-glucopyrano-syloxy)-3-hydroxy-4-octadecene (glucosylsphingosine, **17**, *11*) which via the *N*-4-carboxymethyl-2-nitrobenzyloxycarbonyl derivative **18** was converted to hydrazide **19** and attached to water-soluble polymer **20** (Figure 5). Polymer **20** served as an acceptor in the D-galactosyltransferase reaction (**21**, 35% transfer yield) while further

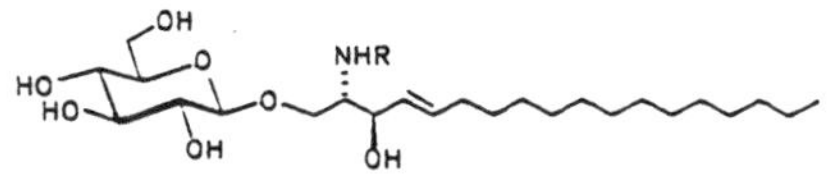

17 R=H
18 R=4-carboxymethyl-2-nitrobenzyloxycarbonyl
19 R=4-hydrazinocarbonyl-2-nitrobenzyloxycarbonyl

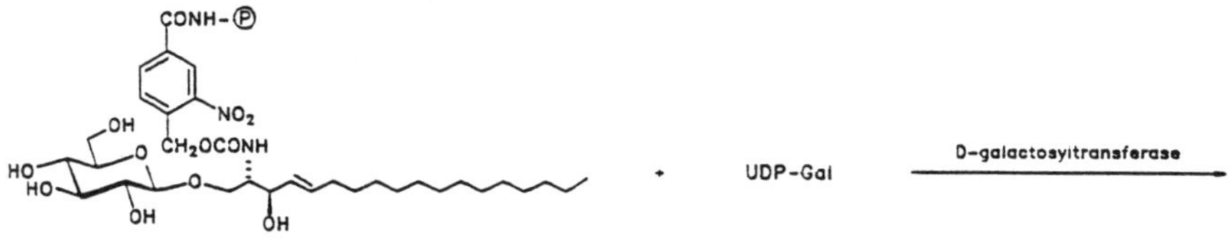

20

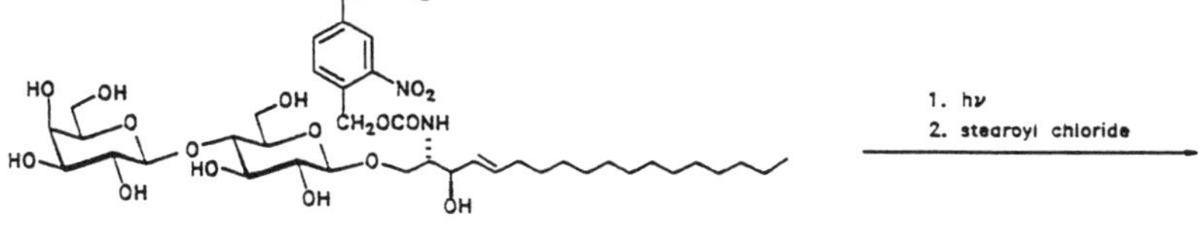

21

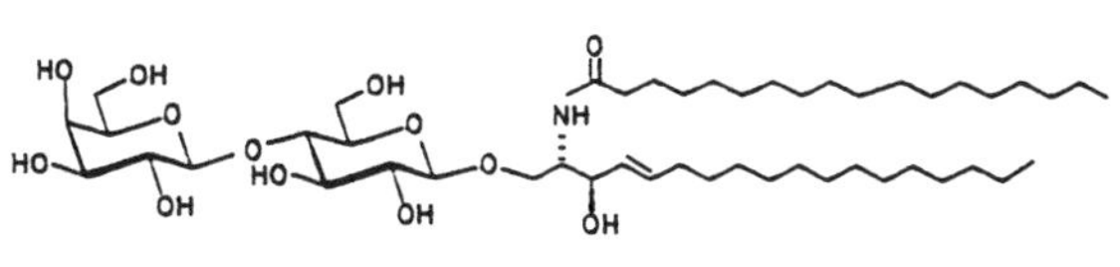

22

Ⓟ = water-soluble substituted poly(acrylamide)

Figure 5. Enzymic synthesis of lactosyl ceramide on a soluble polymeric acceptor.

photolysis, acylation and chromatography afforded $(2S, 3R, 4E)$-1-[4-O-(β-D-galactopyranosyl)-β-D-glucopyranosyloxy]-3-hydroxy-2-octadecanoyl-amino-4-octadecene (lactosyl ceramide **22**, 54% yield) (*12*).

It is assumed that following this first experiment the methodology presented here could be broadened to include additional glycosyl sphingosine-substituted polymers derived from modified native or synthetic precursors which may subsequently serve as acceptors to glycosyltransferases and glycosidases, which takes advantage of the facile isolation of radioactive products while employing the polymeric handle. It may also lead to synthetic *lyso*-glycoshingolipids (free 2-amino functions) and to glycoshpingolipids carrying diverse 2-acylamido substituents.

Conclusion

Hopefully, different variations of the methodology presented here may lead to biologically important oligosaccharides, help establish enzyme specificities, and give access to affinity absorbents for use in isolation of saccharide-binding proteins, such as antibodies, myeloma proteins, lectins, and enzymes.

Acknowledgments

Supported by the Fonds der Chemischen Industrie and by the German-Israeli Foundation for Scientific Research and Development (G.I.F.).

Literature Cited

1. Paulsen, H. *Angew. Chem. Int. Ed. Engl.* **1982**, *21*, 155; Schmidt, R. R. *Angew. Chem. Int. Ed. Engl.* **1986**, *25*, 212; Ogawa, T.; Yamamoto, H.; Nukada, T. *Pure Appl. Chem.* **1984**, *56*, 779.
2. Whitesides, G. M.; Simon, E. S.; Toone, E.; Hung, R.; Borysenko, C.; Grabowski, S.; Spaltenstein, A.; Straub, I; Lees, W. *199th ACS National Meeting*, Boston, MA, April 1990, Abstr. 58; Nilsson, K. G. I. *Ibid*, Abstr. 59; Thiem, J. Abstr. 60; Paulson, J. C. Abstr. 61; Wong, C. -H. Abstr. 62; Hindsgaul, O. Abstr. 67; Nakajima, H.; Kondo, H.; Tsurutani, R.; Dombou, M.; Tomioka, I.; Tomita, K. Abstr. 68; Gygax, D.; Hammel, D.; Schneider, R.; Stierlin, H. Abstr. 69; Mazur, A. W. Abstr. 70; Rasmussen, J. R.; Hirani, S.; McNeilly, D. S.; Chang, M. -Y. G. Abstr. 71.
3. Zehavi, U. *Reactive Polymers* **1987**, *6*, 189.
4. Köpper, S.; Zehavi, U. *5th European Symposium on Carbohydrates*, Prague, Czechoslovakia, August 1989, *Abstr. C-7*.
5. Zehavi, U.; Herchman, M. *Carbohydr. Res.* **1986**, *151*, 371.
6. Zehavi, U.; Herchman, M.; Köpper, S. *199th ACS National Meeting*, Boston, MA, April 1990, Abstr. 47.
7. Treder, W.; Thiem, J.; Schlingmann, M. *Tetrahedron Lett.* **1986**, *27*, 5605.
8. Treder, W.; Zehavi, U.; Thiem, J.; Herchman, M. *Biotechnol. Appl. Biochem.* **1989**, *11*, 362.
9. Thiem, J.; Wiemann, T. *Angew. Chem. Intl. Ed. Engl.* **1990**, *29*, 80.
10. Zehavi, U.; Herchman, M.; Hakomori, S.; Köpper, S. *Glycoconjugate J.* **1990**, *7*, 219.
11. Zimmermann, P.; Bommer, R.; Bär, T.; Schmidt, R. R. *J. Carbohydr. Chem.* **1988**, *7*, 435.
12. Zehavi, U.; Herchman, M.; Schmidt, R. R.; Bär, T. *Glycoconjugate J.* **1990**, *7*, 229.

RECEIVED February 7, 1991

Chapter 8

Galactose Oxidase

Selected Properties and Synthetic Applications

Adam W. Mazur

Miami Valley Laboratories, The Procter and Gamble Company,
P.O. Box 398707, Cincinnati, OH 45239–8707

Galactose oxidase (GOase) catalyzes the oxidation of many primary alcohols, alditols as well as the 6-hydroxy group in D-galactose, D-talose, and D-gulose to the corresponding aldehydes. It is a convenient enzyme for synthetic applications because it can be easily obtained from a fungus and has good stability. GOase is very efficient in the oxidation of a broad range of galactosides, including oligo- and polysaccharides. However, the reactions strongly depend upon the quality of the enzyme preparations. Using a modified isolation method, significant amounts of GOase with the required purity can easily be obtained to enable efficient syntheses, of up to 10 kg, of 5-C-hydroxymethyl-L-arabinohexopyranose derivatives which are not readily available by classical synthetic methods.

This chapter will be devoted to practical aspects of research on galactose oxidase (GOase). Despite its potential to simplify carbohydrate synthesis, this enzyme is not used in synthetic laboratories. The goal of this work is to encourage interested readers to use GOase in their work when such an option exists. Therefore, the material presented in this chapter will emphasize those aspects of our work at Procter and Gamble which, in our perspective, were the most important for the successful development of a practical enzymic oxidation method for galactose-containing carbohydrates.

Our interest in GOase came from a need to find an efficient process for making 5-C-hydroxymethyl-L-arabino-hexopyranose derivatives **1** (Equation 1). These compounds have been developed as candidates for non-nutritive sugar substitutes in foods because they resist metabolism and have sucrose-like functional properties (*1*). These materials would be too expensive if made using the techniques of traditional chemical synthesis, so we developed a simple "one-pot" process (Equation 1) using GOase.

0097–6156/91/0466–0099$06.00/0

Equation 1.   Synthesis of 5-C-Hydroxymethyl-L-arabino-hexopyranoses.

Selected Properties and Applications of Galactose Oxidase

GOase is a copper-containing monomeric metaloprotein, with a molecular
weight of about 68 kDaltons.   It catalyzes the oxidation of many primary
alcohols to the corresponding aldehydes (Equation 2).   The enzyme is very
specific for the oxidation of galactose and its derivatives to galactose 6-
aldehydes (Table I).

$$RCH_2OH + O_2 \longrightarrow RCHO + H_2O_2 \qquad (2)$$

Mechanism.   Since its discovery in 1959 (*2*), galactose oxidase has
attracted considerable interest in the scientific community because of its
enigmatic catalytic mechanism.   Various aspects of early research have
been discussed in two excellent reviews (*3, 4*).   The most puzzling property
of this enzyme has been the ability of the monomeric protein to carry out
two-electron redox reactions without any apparent involvement of a
cofactor.   Although a number of elegant explanations of this phenomena
have been proposed (*3, 5*), the mechanism still remains elusive.   On the
other hand, a recent finding indicates that galactose oxidase does in fact
contain a cofactor (*6*), pyrroloquinolinequinone (PQQ), which is covalently
attached to the protein.   This discovery will certainly add new vitality to the
discussion on its mechanism.   (Note: added in proof.   X-Ray structure at
GOase was recently reported and no presence of cofactor suggested (*33*).)
       Stability.   GOase is stable in buffers at 4°C for weeks, and
indefinitely when quickly frozen and kept at -70 C.   However, it loses 75%
of its activity upon lyophilization.   Having copper as an essential
component at the active center, GOase is inactivated by ions reactive with
copper such as sulfide, cyanide, azide, and many copper chelators.
Hydrogen peroxide (a product of the oxidation reaction) is also a potent
deactivator and must be decomposed immediately *in situ.*   This decompostion
is usually accomplished by running the reaction in the presence of
catalase.
       Analytical Applications.   GOase has been used mainly as an
analytical reagent.   It is commercially available as "Galactostat"™
(manufactured by Worthington Biochemicals) and used for measuring
galactose and its derivatives in biological fluids.   GOase also has been used to
label galactose residues in glycoconjugates.   The procedure involves a
desialylation step with neuramidase followed by the oxidation of the
exposed galactosyl residues with galactose oxidase to the corresponding 6-
aldehydes.   The carbonyl group is subsequently transformed to a
functionality with easily detectable substituents (*7*) such as tritium,
deuterium, or fluorine, carboxyl, amine, nitroxide spin labels and various
chromophores.
       Among more recent applications, Kiba and coworkers (*8*) have built
a post-column HPLC detector, containing co-immobilized galactose oxidase
and peroxidase to detect stachyose, raffinose, mellibiose and galactose in
trace amounts.   Another interesting application involves an integrated
biosensor for simultaneous but independent detection of galactose and
glucose (*9*).   Such analysis is possible because glucose dehydrogenase
(GDH), used as the glucose sensor, uses NAD^+ as an electron acceptor
whereas GOase uses oxygen;   these acceptors do not exchange electrons.
       Synthetic Applications.   There have been only two reports
describing synthetic applications of GOase: Yalpani and Hall (*10*) described

Table I.   Relative   reactivities   of   galactose   oxidase   with   carbohydrates
(Refs. 3, 30, 31.)

Substrate	Relative Reactivity	Substrate	Relative Reactivity
D-Galactose	100		
β-Methylgatactopyranoside	340	D-Talose	52
3-0-Methylgalactose	176	D-Gulose	0.08
2-0-Methylgalactose	50	D-Glucose	0
4-0-Methylgalactose	0	L-Galactose	0
Stachyose	610		
Raffinose	180		
Lactose	2		

modifications of galactomannans while Wong and coworkers (*11*) reported the synthesis of some uncommon sugars.   The potential of GOase is certainly much larger than currently explored.   The successful development of galactose oxidase as a synthetic reagent would make possible many selective transformations without the need for protecting groups.   This could certainly lead to more extensive use of inexpensive, safe, and biodegradable carbohydrate materials in industrial processes.

In our view, a large barrier in the development of GOase as a practical synthetic reagent has been the limited availability of sufficiently pure enzyme.   The commercial preparations are very expensive and only 10 to 20% of the total protein possesses GOase activity.   Most of these products are obtained by lyophilization, a process known to lower the activity by at least 75%.

A separate problem has been the considerable sensitivity of GOase to inhibition and inactivation by macromolecular contaminants. Unfortunately, this may be a general property of the enzyme, and not limited to specific inhibitors such as those present in fermentation broths. A likely factor responsible for these phenomena may be that the high basicity of the enzyme (p*I* 12) promotes strong interactions between the protein and other macromolecular assemblies.   For example, the enzyme is known to adhere to quartz and glass so strongly that it can only be completely removed by washing the surfaces with an acid solution (*3*). Moreover, the oxidations of terminal galactosyl or *N*-acetylgalactosaminyl residues of glycoconjugates are almost never complete and become even less effective when the substrates are attached to the cell membranes (*7*). We also noticed similar sensitivity to contaminants when we tried to use crude preparations of GOase to oxidize a simple sugar, methyl D-galactoside. In this case, only 40% of the substrate was converted to the corresponding aldehyde, under the standard conditions.   Recovery of the enzyme by ultrafiltration did not improve its properties: when added to a new batch of methyl D-galactoside, the enzyme produced only 8% of the aldehyde.   On the other hand, the full activity and normal kinetic properties were restored when, after recovery from the reaction mixture, the enzyme was purified on a carboxysulfone column, according to the procedure described in the next section.   The described phenomena indicate that the catalytic activity of GOase may be modified by intermolecular interactions with macromolecules.   Consequently, we do not recommend the use of crude GOase preparations in synthesis because it may lead to ambiguous results.

Preparation of GOase

Galactose oxidase is produced by a number of fungal species (*12-14*) but is usually obtained from *Dactylium dendroides* (*15*).   Although initially the enzyme was described as being extracellular, an intracellular form also was recently discovered (*16, 17*).   A traditional purification method, based on precipitation in the presence of microcrystalline cellulose, was described by Horecker and coworkers (*18, 19*).   Later, Kosman and Tressel (*15*) introduced chromatographic purification on phosphocellulose as the final step.   They recommended two passes through the column to remove completely a second protein, which strongly inhibits galactose oxidase (*11*). Other chromatographic versions of the final purification are also available.   One of them described for commercial samples involves separation on Sepharose-6B (*20*) and another, more recent, is an affinity procedure using a melibiose-polyacrylamide column (*21*).

We prepared galactose oxidase in 30-L and 250-L fermentation batches using a strain of *Dactylium dendroides*. The 250-L fermentation batches were prepared at Lehigh University. The fermentation is aerobic and takes 5 to 7 days. A typical plot of the enzyme activity in the medium versus time is shown in Figure 1. Initially, we used the isolation method of the extracellular enzyme according to Kosman and Tressel (*15*) and obtained a good quality enzyme. Unfortunately, the procedure was very laborious, requiring 3 days of work in the cold room per 30 L fermentation volume. Furthermore, we encountered substantial technical problems while trying to develop the process for 250-L batches, so we developed a new isolation method (Figure 2).

Improved Isolation Procedure. Our improved method is based on ultrafiltration of the supernatant followed by purification of the crude enzyme on a carboxysulfone ion-exchange column. Treatment with DEAE cellulose, done after initial concentration of the supernatant, is optional but useful: it removes about half of the concomitant proteins. In the final step, the crude solution is loaded in portions on the ion-exchange column and the void volumes are isocratically eluted after each injection with the 4-morpholine ethanesulfonic acid (MES) buffer. Finally, sodium acetate gradient is imposed to elute the enzyme. Using a 40 micron BAKERBOND CARBOXYSULFONE™ column (4.6 x 250 mm) from J. T. Baker Co., several hundred milligrams of GOase can be purified in a single elution. The galactose oxidase-inhibiting protein (*15*) has a substantially longer retention time under these conditions and there is no need for the second pass. With the proper ultrafiltration equipment and trained personnel, the entire procedure takes 1 to 1.5 days and can give 70% yield of recovered activity.

Effects of Copper. We found that an important factor for maintaining GOase activity during fermentation, purification and its use in synthesis was the presence of copper ions in the enzyme solutions. Increases in Cu^{2+} from 11 to 55 μmol/L improved the reproducibility of the fermentations, but higher concentrations inhibited fungal growth. During ultrafiltration, and HPLC purification, 2-20 mM Cu^{2+} concentration had to be maintained, otherwise 25 to 50% of GOase activity was lost irreversibly in each step. Avigad and Markus (*22*) argued that deactivation in crude samples is caused by GOase binding to heptapeptide (MW 724), and it can be prevented by the presence of copper ions. However, in our case, the main factor responsible for the inhibition was apparently different since filtration through 30,000 MWCO membrane did not slow down the decrease in activity in the absence of Cu^{2+}. In contrary, the inactivation rate was proportional to the increase in total concentration of the retentate. Thus, the inactivation could not be caused by a small molecule. Furthermore, our synthetic work indicates that the stabilizing effect of copper is not limited to the impure samples since the presence of Cu^{2+} in the reaction mixture, during the oxidations of carbohydrates, also improved the stability of the purified enzyme (Figure 3).

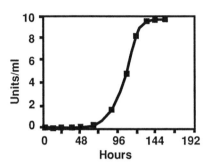

Figure 1. Enzyme activity in fermentation medium during production of galactose oxidase (HRP activity assay).

FERMENTATION BROTH, 30L

   ↓ Mycelia filtration

SUPERNATANT(250-300 U/mg)

   ↓ Ultrafiltration, Pellicon system, 10,000 MWCO

CONCENTRATED ENZYME, 300-500 mL

   ↓ DEAE batch absorption

   ↓ Ultrafiltration, Amicon cell

CONCENTRATED ENZYME 15 mL

   ↓ HPLC (carboxysulfone column)

ENZYME (1500- 2200 U/mg)

Figure 2. Improved process for the isolation of galactose oxidase (HRP activity assay).

Assay Procedures. Three methods are available for determining galactose oxidase activity. A coupled assay employs horseradish peroxidase (HRP) which catalyses a hydroxylation by hydrogen peroxide of an aromatic chromogenic substrate, such as o-dianisidine. The activity units, in this assay, are defined as the rate of absorbance increase. On the other hand, two other assays directly measure changes in the concentration of the reactant or product; oxygen consumption using the Clark-type electrode, or the production of aldehyde. In general, the most sensitive is the chromogenic assay and it is routinely used during the enzyme preparation. The O_2 uptake is the most convenient method for determining the initial rates due to the fast response of the electrode to the changes in oxygen concentration. The direct oxidation assay can be conveniently tailored to the actual needs. In particular, one can use aromatic alcohols and follow the reaction by monitoring the changes in UV absorbance, or employ a characteristic reaction for the aldehyde group. The latter option is the best for recording the progress of synthetic reactions. A discussion of various conditions effecting GOase assays can be found in papers by Hamilton and coworkers (23, 24).

Practical Synthesis Using GOase

General conditions for the oxidations of galactose-containing carbohydrates are as follows (Figure 4): the reactions were run at 4°C to minimize the deactivation of GOase, inhibit microbial growth and maximize the solubility of oxygen. Vigorous aeration and stirring were required in order to provide effective transport of oxygen and hydrogen peroxide to the system. In the batches where the air flow or stirring were interrupted, the oxidation stopped and a substantial amount of enzyme was irreversibly deactivated. Finally, as was already mentioned the presence of Cu^{2+} (CuSO4, 0.5 mM) was beneficial for enzyme stability.

Figure 5 lists the aldehydes which were made in quantities ranging from 10 g to 10 kg. Figure 6 shown a representative NMR spectrum of the aldehyde taken from the unpurified residue after lypophilization of the reaction mixture. This spectrum illustrates the specificity of the enzyme. Initially, we were concerned about the possible formation of uronic acids reported by some authors (25-27). However, we have no evidence for the presence of these by-products from the carbon-13 spectra of the oxidation products. On the other hand, the small signals between 90 ppm and 103 ppm may belong to dimeric products described by Maradufu and Perlin (28). All the aldehydes are hydrated as indicated by the presence of a peak at 88.7 ppm and the absence of peaks below 200 ppm. These aldehydes are relatively stable in the presence of base as evidenced by preparation of 5-C-hydroxymethyl-L-arabino-hexopyranoses (Equation 1). Even though the synthesis was carried out in 1 N NaOH, very little by-products resulted from elimination or retroaldol reactions at the pyranose ring.

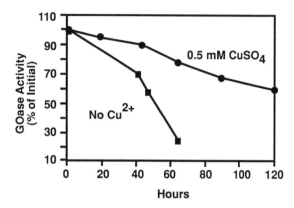

Figure 3. Effect of Cu^{2+} on GOase stability during oxidation of methyl α-D-galactoside.

Synthetic Conditions	
Carbohydrate,	0.25 - 0.600 M;
phosphate buffer,	50 mM (pH 7.0);
galactose oxidase,	3600 U/100 mmol substrate;
catalase,	10 U/1 U GOase;
temperature,	4°C;
$CuSO_4$,	0.5 mM;
The reaction mixture is stirred and aerated	
Yields are usually >80%.	

Figure 4. Conditions for the oxidations of galactose-containing carbohydrates.

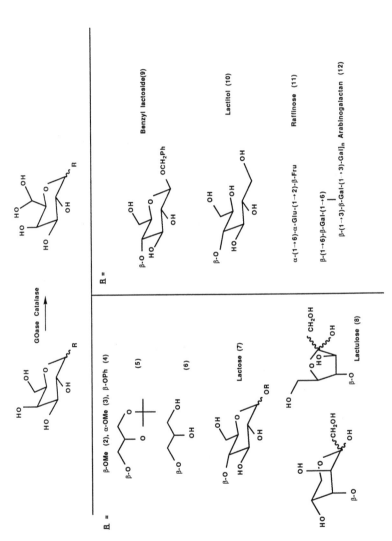

Figure 5. Aldehydes prepared in 10 g to 10 kg quantities using galactose oxidase.

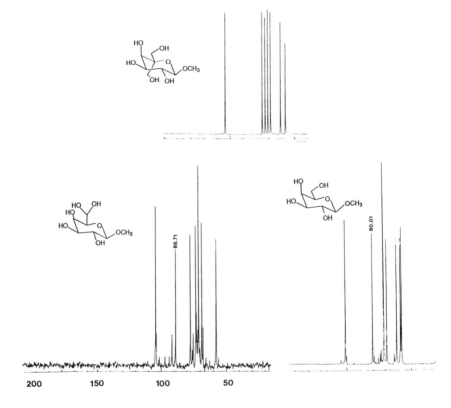

Figure 6. Representative carbon NMR spectra of substrate and products for the reaction in Equation 1

Literature   Cited

1.   Eur. Patent Applications 89304505.4, and 89304504.7, Nov. 89.
2.   Copper, J. A. D.; Smith, W.; Bacila, M.; Medina, H. *J. Biol. Chem.* **1959**, *234*, 445-448.
3.   Ettinger, M. J.; Kosman, D. J. In *Copper Proteins*; Spiro, T. G., Ed.; John Wiley & Sons: New York, 1981; Chapter 6, pp 221-261.
4.   Kosman, D. J. In *Copper Proteins and Copper Enzymes*; Lontie, R., Ed.; CRC Press, 1984; Vol. 2, Chapter 1, pp 2-26.
5.   Whittaker, M. M.; Whittaker, J. W. *J. Biol. Chem.* **1988**, *263(13)*, 6074-6080.

6.   van der Meer, R. A.; Jongejan, J. A. *J. Biol. Chem.* **1989**, *264(14)*, 7792-7794.
7.   Avigad, G. *Arch. Biochem. Biophys.* **1985**, *239(2)*, 531-537.
8.   Kiba, N.; Shitara, K.; Furosawa, M. *J. Chromatogr.* **1989**, *463*, 183-187.
9.   Yokoyama, K.; Sode, K.; Tamiya, E.; Karube, I. *Anal. Chim. Acta* **1989**, *218(1)*, 137-142.
10. Yalpani, M.; Hall, L. D. *J. Polym. Sci: Chem. Ed.* **1982**, *20*, 3399-3420.
11. Root, R. L.; Durrwachter, J. R.; Wong, C. J. *Am. Chem. Soc.* **1985**, *197*, 2997-2999.
12. Gancedo, J. M.; Gancedo, C.; Asensio, C. *Arch. Biochem. Biophys.* **1967**, *119*, 588-590.
13. Kazuo, A.; Terado, O. *Agric. Biol. Chem.* **1981**, *45(10)*, 2311-2316.
14. Koroleva, O. V.; Rabinovich, M. L.; Buglova, T. T.; Yaropolov, A. I. *Prikl. Biokhim. Microbiol.* (translation) **1983**, *19(5)*, 632-637.
15. Tressel, P.; Kosman, D. J. *Anal. Biochem.* **1980**, *105*, 150-153.
16. Medonca, M. H.; Zacan, G. T. *Arch. Biochem. Biophys.* **1987**, *252(2)*, 507-514.
17. Medonca, M. H.; Zacan G. T. *Arch. Biochem. Biophys.* **1988**, *266(2)*, 427-434.
18. Amaral, D.; Bernstein, L.; Morse, D.; Horecker, B. L. *J. Biol. Chem.* **1963**, *238*, 2281-2284.
19. Amaral, D.; Kelly-Falcoz, F.; Horecker, B.L. *Methods Enzymol.* **1966**, *9*, 87-92.
20. Hatton, M. W. C.; Regoeczi, E. *Methods Enzymol.* **1982**, *89*, 172-176.
21. Kelleher, F. M.; Dubbs, S. B.; Bhavanandan, V. P. *Arch. Biochem. Biophys.* **1988**, *263(2)*, 349-354.
22. Avigad, G.; Markus, Z. *Fed. Proc.* **1972**, *31*, 447.
23. Hamilton, G. A.; Adolf, P. K.; DeJersey, J.; DeBois, G. C.; Dyrkacz, G. R.; Libby, R. D. *J. Am. Chem. Soc.* **1978**, *100*, 1899-1912.
24. Hamilton, G. A.; DeJersey, J.; Adolf, P. K. In *Oxidases and Related Redox System*; King, T. E.; Mason, H. S.; Morrison, M., Eds.; University Park Press: Baltimore, MD; 1973; pp 103-124.
25. Pazur, J. H.; Knull, H. R.; Chevalier, G. E. *J. Carbohydr. Nucleoside Nucleotides* **1974**, *4(2)*, 129-146.
26. Matsumura, S. et al. *Chem. Lett.* **1988**, 1747-1750.
27. Matsumura, S.; Kuroda, A.; Higaki, N.; Hiruta, Y.; Yoshikawa, S. *Chem. Lett.* **1988**, 1747-1750.
28. Maradufu, A.; Perlin, A. S. *Carbohydr. Res.* **1974**, *32*, 127-136.
29. Kelleher, F. M. and Bhavanandan, V. P. *J. Biol. Chem.* **1986**, *261(24)*, 11045-11048.
30. Avigad, G.; Amaral, D.; Asensio, C.; Horecker, B. L. *J. Biol. Chem.* **1962**, *237(9)*, 2736-2743.
31. Schlegel, R. A.; Gerbeck, C. M.; Montgomery, R. *Carbohydr. Res.* **1968**, 7, 193-199.
32. Godtfredsen, S. E. et al. In *Enzymes as Catalyst in Organic Synthesis*; Schneider, M. P., Ed.; NATO ASI Series Vol. 178; D. Reidel Publishing, 1985; pp 77-95.
33. Ito, N. et al. *Biochem. Soc. Trans.* **1990**, *18(5)*, 931-932.

RECEIVED February 7, 1991

Chapter 9

Industrial Application of Adenosine 5'-Triphosphate Regeneration to Synthesis of Sugar Phosphates

H. Nakajima, H. Kondo, R. Tsurutani, M. Dombou, I. Tomioka, and K. Tomita

Department of Biochemistry, Research and Development Center, Unitika Ltd., 23 Kozakura, Uji, Kyoto 611, Japan

Enzyme-catalyzed reactions are attracting attention in the field of chemical synthesis, especially for the production of fine chemicals such as pharmaceuticals and agricultural chemicals. Although many studies concerning bioreactor systems employing enzymes as catalysts have been intensively investigated, most of the investigations are still at the stage of fundamental research. Consequently, there is an increasing demand for demonstrations that bioreactors can be employed for the production of fine chemicals on an industrial scale. For this purpose, we have developed purification methods for many types of thermostable enzymes and have used them to produce valuable sugar phosphates on a kilogram scale. We also have demonstrated the preparation of peptides and a fascinating nucleotide, diadenosine tetraphosphate, using aminoacyl tRNA synthetase and an ATP regeneration system.

Most biosynthetic reactions used in industry employ hydrolytic enzymes. Many other enzymes could be used in synthesis, but these enzymes require adenosine 5'-triphosphate (ATP) as a source of energy. The synthesis of various fine chemicals using bioreactors with systems for the regeneration of ATP has been investigated intensively on a research scale. We have produced sugar phosphates in kilogram-scale reactions to demonstrate that bioreactors can be used to produce fine chemicals on an industrial scale.

The use of enzymes in bioreactors is expected to be an ideal process for the chemical industry. Enzymes, however, have serious disadvantages from an industrial point of view: most preparations of enzymes are unstable because they were derived from mesophiles. It is already well known that the enzymes derived from thermophillic organisms, thermophiles, are thermally stable and resistant to organic solvents and surfactants. In our work we have used thermostable enzymes to impart durability to bioreactors.

0097–6156/91/0466–0111$06.00/0

In this report, we describe the general properties of thermostable enzymes, an ATP regeneration system using thermostable enzymes, and four applications of our thermostable ATP regeneration system.

Thermostable Enzymes

Microorganisms are classified into three groups based on their maximum growth temperatures (1): psychrophiles (<30°C), mesophiles (30-55°C), and thermophiles. Thermophiles, which usually grow at temperatures higher than 55°C such as in hot springs, can be classified into two subdivisions; moderate thermophiles (55-75°C) and extreme thermophiles (>75°C). Most thermostable enzymes are obtained from thermophiles, although some, such as thermostable protease, are from mesophiles. We have purified 20 types of thermostable enzymes from the thermophile, *Bacillus stearothermophilus*.

Figure 1 shows that enzymes from thermophiles are more thermostable than enzymes isolated from mesophiles. For example, acetate kinase (AK, E.C. 2.7.2.1) from *B. stearothermophilus* retained its activity until 60°C. Thermostable enzymes are especially stable at or below room temperature (Figure 2). As a result enzymes from thermophiles are useful in applications requiring stability or long shelf life (2). Now several thermostable enzymes from thermophiles are commercially available, such as thermolysin from *B. thermoproteolytics*, restriction enzymes and DNA polymerase from *Thermus aquaticus* and *T. thermophilus*, enzymes for the regeneration of ATP, glycolytic enzymes and superoxide dismutase from *B. stearothermophilus*.

ATP Regeneration System

The various possibilities for large-scale regeneration of ATP which have been investigated so far can be classified into five methods (3): extraction, chemical synthesis, use of intact or disrupted cells, use of organelles, and use of cell-free enzymes. The use of cell-free enzymes provides the simplest and most feasible route to ATP regeneration on an industrial scale. The use of cell-free enzymes has many advantages, including a high degree of specificity, absence of undesirable by-products, and potentially high conversion of reactants to products. These advantages make it worthwhile to overcome problems such as the purification of enzymes and the preparation of phosphoryl donors. Furthermore, advances in techniques for the immobilization and stabilization of enzymes have greatly enhanced their potential use as highly specific catalysts in large-scale industrial processes.

We have examined the use of AK for the regeneration of ATP on an industrial scale. We have developed a regeneration system using thermostable AK and thermostable adenylate kinase (AdK, E.C. 2.7.4.3) to regenerate ATP from ADP or AMP. Two reactions convert AMP to ATP. In the first reaction, AdK converts AMP and ATP to ADP (Equation 1). AK converts ADP to ATP in the second reaction, using acetyl phosphate (AcOP) as the phosphoryl donor (Equation 2). By combining these two reactions, it is evident that two equivalents of AcOP can convert one equivalent of AMP to ATP (Equation 3).

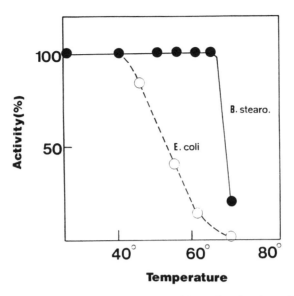

Figure 1.   Comparison of the thermostability of adenylate kinase from a mesophile (*E. coli*) and from a thermophile (*B. stearo.*).   Solutions of adenylate kinase were incubated at various temperatures for 15 min and then assayed for remaining activity.

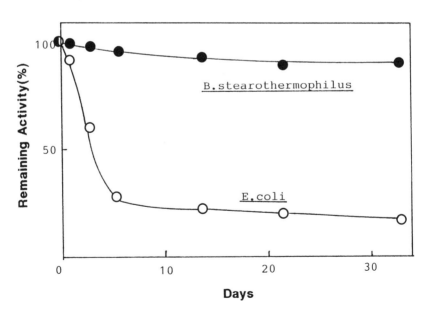

Figure 2.   Comparison of the storage stability at room temperature of adenylate kinase from a mesophile (*E. coli*) and from a thermophile (*B. stearothermophillus*).

$$\text{AMP} + \text{ATP} \xrightarrow{\text{AdK}} 2\,\text{ADP} \qquad (1)$$

$$2\,\text{ADP} + 2\,\text{AcOP} \xrightarrow{\text{AK}} 2\,\text{ATP} \qquad (2)$$

$$\text{AMP} + 2\,\text{AcOP} \qquad\qquad \text{ATP} + 2\,\text{AcOH} \qquad (3)$$

Our ATP regeneration system $(4, 5)$ is composed of the following five units (Figure 3): 1) a substrate unit (containing AMP, ATP and AcOP), 2) an enzymatic reactor unit, 3) an auto sampler unit, 4) an analytical unit, and 5) a control unit. Although AcOP easily degrades, it can be used for long-term operation of the ATP regeneration system if kept at around 0°C. The enzymatic reactor is a column type, in which AK and AdK are co-immobilized. We adapted the autosampler unit of an HPLC apparatus to our bioreactor by changing the sampling mechanism, and we used the HPLC apparatus as the analytical unit. The control unit is a microcomputer. Based on analysis of the concentrations of AMP, ADP and ATP, the control unit directs the motion of the pumps and the electric valves through the electric relay systems and consequently the concentrations and the flow rates of the substrates can be regulated to keep the optimum operational conditions for ATP regeneration. Even after 6 days, the extent of conversion of AMP to ATP was at 100%. We can operate the reactor continuously for longer periods of time.

Production of Pharmaceutical Intermediates

There are many reports (6) of the production of various kinds of valuable materials in bioreactors with coupled regeneration of ATP (glutathione, CDP-choline, Coenzyme A, glutamine, gramicidin S, NADP, glucose 6-phosphate, glycerol 3-phosphate, phosphoribosyl 1-pyrophosphate, creatine phosphate). Most of these investigations are still at the stage of basic research. Consequently, there is an increasing demand for proof that bioreactors can be used to produce fine chemicals on an industrial scale. As examples, we examined the production of glucose 6-phosphate (G-6-P) and fructose 1,6-diphosphate (FDP) on a scale of several hundred kilograms.

Synthesis of Glucose 6-Phosphate. G-6-P is a useful reagent for enzyme immunoassays (EIA), such as Syva's EMIT system, for NADH regeneration, and as an intermediate for the synthesis of oligosaccharides. Glucose is converted to G-6-P accompanied by the conversion of ATP to ADP in the presence of thermostable glucokinase (GlcK, E.C. 2.7.1.2). ADP thus formed is easily regenerated into ATP by the action of thermostable AK in the presence of AcOP (Figure 4).

We now produce G-6-P continuously in a flow reactor which is packed with immobilized thermostable AK and GlcK (7). Using this reactor system, we have demonstrated the continuous production of G-6-P for up to 40 h. Since ATP is regenerated in situ, only 1mM of ATP, or one percent equivalent of glucose, is enough to fully convert glucose to G-6-P continuously.

Synthesis of Fructose 1,6-Diphosphate. Production of FDP is another example of the use of our ATP regeneration system. FDP is used as an intravenous therapeutic agent. It is used in resuscitation, total parenteral nutrition, and cardiology. An Italian company produces more than 20 tons of FDP per year by fermentation. The fermentation process

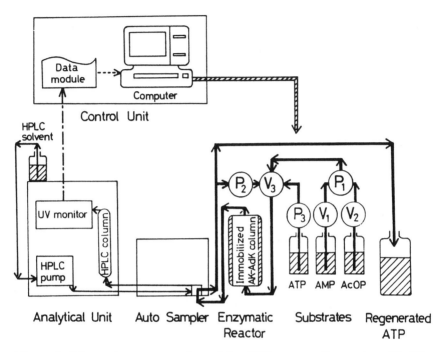

Figure 3. Schematic representation of the ATP regeneration system. P1, P2, P3 and V1, V2, V3 represent peristaltic pumps and electric valves, respectively.

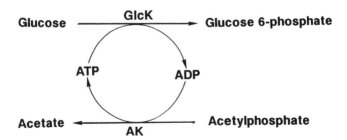

Figure 4.   Synthesis of G-6-P using a bioreactor coupled with ATP regeneration system.   GlcK = glucokinase; AD = adenylate kinase.

suffers from low percent of conversion and the laborius use of columns for the isolation of product.

We can produce F-6-P and FDP from fructose in kilogram-quantities. The catalytic action of hexokinase (E.C. 2.7.1.1), coupled to an ATP regeneration system, converts fructose to fructose 6-phosphate. The action of phosphofructokinase (E.C. 2.7.1.11), again coupled to ATP regeneration, converts fructose 6-phosphate to FDP.

Glucose is much less expensive that fructose, however, so we tried to synthesize FDP from glucose using a different system involving GlcK, phosphoglucose isomerase (E.C. 5.3.1.9), and phosphofructokinase (Figure 5). Although the equilibrium constant for the phosphoglucose isomerase reaction is nearly one, the reaction catalyzed by phosphofructokinase drives the process to completion.

We have prepared FDP using crude whole cell extracts of *B. stearothermophilus* (8). The extract from *B. stearothermophilus* contains the enzymes required for the main reaction and for the regeneration of ATP. The fact that we used a crude cell extract to completely convert glucose to FDP is significant from an industrial point of view because a crude cell extract is less expensive than purified enzymes. A crude extract from yeast did not produce FDP under similar conditions. We are now studying the optimal conditions and mechanism of this process.

Synthesis using Aminoacyl t-RNA Synthetase. Other interesting synthetic reactions using ATP are the use of aminoacyl t-RNA synthetase (ARS, E.C. sub-class 6.1.1) to synthesize peptides (9-12) and for the construction of the fascinating biologically-active nucleotide, diadenosine polyphosphate (Ap4A) (13-14). ARS isolated from mesophiles is quite unstable and is usually handled on ice. ARS from *B. stearothermophilus*, on the other hand, is quite stable and is a useful catalyst for bioreactors.

Peptides. It is well know that, in the first stage of aminoacyl t-RNA formation, an amino acid reacts with ATP, and aminoacyl AMP intermediate is formed. This intermediate is easily attacked by nucleophiles such as the amino group of a second amino acid or peptide (Equation 4). Table I summarizes some of the dipeptides produced using this reaction. The reac-

$$AA_1 + AA_2 + ATP \xrightarrow{\text{ARS}} AA_1\text{-}AA_2 + AMP + PP_i \qquad (4)$$

tion is important because a variety of D- and L-amino acid derivatives with and without protecting groups can be used an nucleophiles and because it is possible to introduce proline in peptide bonds, which is impossible in the conventional enzymatic synthesis of peptides using proteases. Furthermore, this reaction is carried out under mild conditions in aqueous solution and no racemization takes place. We applied this reaction to prepare a labeled opioid peptide, Kyotorphine (12), using immobilized tyrosyl t-RNA synthetase (TyrRS)

P, P'-Diadenosine tetraphosphate (Ap4A). Ap4A (Figure 6). has various biological activities such stimulating DNA synthesis, inhibiting protein kinases, adenylate kinase, platelet aggregation, and signaling the oxidation or heat stress of cells (15). The preparation of this compound by chemical synthesis has been reported (16). In the chemical synthesis, AMP is activated as a morpholidate or diphenyl phosphate. These compounds

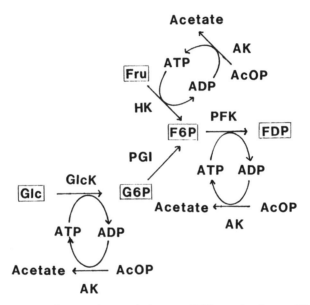

Figure 5. Enzymatic reactions relating to FDP production. Glc = glucose; G6P = glucose 6-phosphate; Fru = fructose; F6P = fructose 6-phosphate; FDP = fructose 1,6-diphosphate; AcOP = acetyl phosphate; AK = acetate kinase; GlcK = glucokinase; HK = hexokinase; PFK = phosphofructokinase.

Table I. Dipeptides Prepared Using Aminoacyl-tRNA Synthetase according to Equation 4

AA$_1$	AA$_2$	Yield[a] (%) of AA$_1$-AA$_2$
L-Tyr	L-LeuOBu^t	50
L-Tyr	LeuNH$_2$	70
L-Tyr	L-PheNH$_2$	41
L-Tyr	L-ProNH$_2$	79
L-Tyr	L-ArgOH	61[b]
L-Tyr	D-LeuNH$_2$	68
D-Tyr	L-LeuNH$_2$	72
L-Leu	L-PheNH$_2$	77
L-Leu	L-PheNH$_2$	82
L-Leu	L-ProNH$_2$	83
L-His	L-ProNH$_2$	87
L-Asp	L-LeuNH$_2$	87

[a] Yield is based on the the amount of AA$_1$. A reaction mixture (volume of 150 μL) containing 180 μM ^{14}C-AA$_1$, 5 mM ATP, 5 mM MgCl$_2$, 50 mM Bicine buffer (pH 8.5), 100 mM AA$_2$ and approximately 4 μM ARS was incubated at pH 8.5 and 30°C for 2 days. The product was separated using paper chromatography and the amount of radioactivity was quantified using a liquid scintillation counter.

[b] A reaction mixture (volume of 1 mL) containing 8 mM AA$_1$, 20 mM ATP, 20 mM MgCl$_2$, 50 mM Bicine buffer (pH 8.5), 100 mM AA$_2$ and 2.5 μM ARS was incubated at 45°C for 2 days. The formation of product was monitored by HPLC.

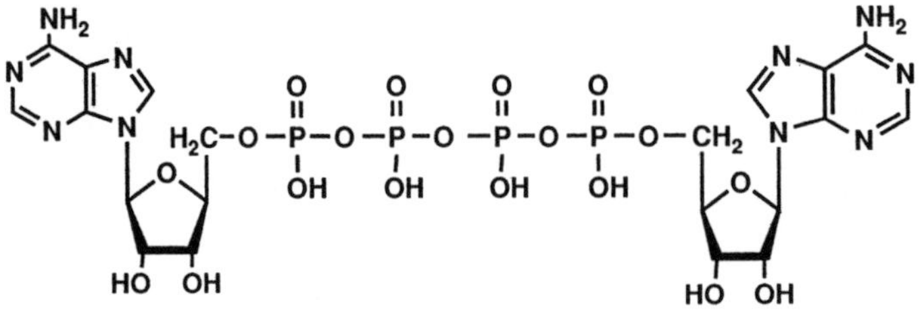

Figure 6. Structure of diadenosine tetraphosphate (Ap4A).

react with ATP to from Ap4A. The reactions have low yields (about 25%), and have many by-products.

Ap4A is produced *in vivo* using ARS. The reaction is not simple. When ATP was incubated with leucyl tRNA synthetase, various by-products, such as Ap3A, ADP and AMP were produced in addition to Ap4A. AMP and ADP were generated by chemical and enzymatic hydrolysis of ATP. Ap3A is produced enzymatically from ADP and ATP. By coupling our ATP regeneration system using AK and AdK with leucyl tRNA synthetase, the formation of by-products was suppressed. This method provides a practical route for the synthesis of this biologically important compound.

Both peptide and Ap4A synthesis reactions are carried out at 40°C. After one day, more than 80% of original activity of ARS remained. The disadvantage of these reactions using ARS may be the fairly low content of ARS in the cell. Hopeful findings were reported by Barker, however, who cloned TyrRS gene of *B. Stearothermophilus* into *E. coli*, thereby increasing the content of TyrRS by a factor of 100 more than in the original cell (*17*). We are also cloning the ARS gene into *E. coli*.

Conclusion

We have established a process for the continual regeneration of ATP from ADP or AMP, applicable to industrial use. The use of thermostable enzymes imparts durability to the regeneration system. We have used our ATP regeneration system to produce hundreds of kilograms of important sugar phosphates on a large scale, and also to produce peptides and Ap4A using aminoacyl t-RNA synthetase.

Acknowledgments

We are very grateful to Dr. K. Imahori and Dr. K. Suzuki for leading these works and for encouraging us.

Literature Cited

1.   Ohshima, T. "Thermophilic Bacteria"; Tokyo University Shuppan, 1978.
2.   UNITIKA ENZYME Catalogue.
3.   Langer, R. S.; Hamilton, B. K.; Garder, C. R.; Archer, M. C.; Colton, C. K. *AIChE J.* **1976**, *22*, 1079.
4.   Nakajima, H.; Nagata, K.; Kondo, H.; Imahori, K. *J. Appl. Biochem.* **1984**, *6*, 19.
5   Kondo, H.; Tomioka, I.; Nakajima, H.; Imahori, K. *J. Appl. Biochem.* **1984**, *6*, 29.
6.   Chenault, H. K.; Simon, E. S.; Whitesides, G. M. In *Genetic Engineering and Biotechnology Reviews*; Russell, G. E., Ed.; Intercept Ltd.; Newcastle upon Tyne, 1988; Vol. 6, Chapter 6.
7.   Nakajima, H.; Tomioka, I.; Kitabatake, S.; Tomita, K. *Nippon Dempun Gakkai Yoshisyu* **1986**, 218.
8.   Tsurutani, R.; Onda, M.; Ishihara, H.; Nakajima, H.; Tomita, K. *Nippon Nogeikagaku Kaishi* **1989**, *63*, 748.
9.   Nakajima, H.; Kitabatake, S.; Tsurutani, R.; Tomioka, I.; Yamamoto, K.; Imahori, K. *Biochim. Biophys. Acta.* **1984**, *790*, 197.

10. Nakajima, H.; Kitabatake, S.; Tsurutani, R.; Yamamoto, K.; Tomioka, I.; Imahori, K. *Int. J. Prep. Prot. Res.* **1986**, *28*, 179.
11. Dombou, M.; Nakajima, H.; Kitabatake, S.; Tomioka, K.; Imahori, K. *Agric. Biol. Chem.* **1986**, *50*, 2967.
12. Kotabatake, S.; Tsurutani, R.; Nakajima, H.; Tomita, K.; Yoshihara, Y.; Ueda, H.; Takagi, H.; Imahori, K. *Pharmaceutical Res.* **1984**, *4*, 154.
13. Kitabatake, S.; Dombou, M.; Tomioka, I.; Nakajima, H. *Biochem. Biophys. Res. Commun.* **1987**, *146*, 173.
14. Nakajima, H.; Tomioka, I.; Kitabatake, S.; Dombou, M.; Tomita, K. *Agric. Biol. Chem.* **1989**, *53*, 615.
15. Zamecknik, P. *Anal. Biochem.* **1983**, *134*, 1.
16. Reiss, J. R.; Moffatt, J. G. *J. Org. Chem.* **1965**, *30*, 3381.
17. Barker, D. G. *Eur. J. Biochem.* **1982**, *125*, 357.

RECEIVED February 1, 1991

APPENDIX

Classification of Enzymes Referred to in This Volume[1]

Aldolases and Synthetases

Aldolase	1, 2
Fructose-1,6-diphosphate aldolase	2, 23, 24, 28
2-Deoxyribose-5-phosphate aldolase	23, 28
N-Acetylneuraminic acid aldolase	8, 14, 23, 32
Fuculose 1-Phosphate Aldolase	4
KDO synthetase	8
Tyrosyl t-RNA synthetase	116
Leucyl tRNA synthetase	119
Aminoacyl tRNA synthetase	111, 116, 119
Transketolase	2, 8
Transaldolases	8

Glycosidases

N-Acetyl-β-galactosaminidase	54, 55
N-Acetyl-β-glucosaminidase	54, 55, 57, 58
N-Acetylglucosaminidase	53
Endoglycosidase H	58, 84
Endo-α-N-acetylgalactosaminidase	58
Endo-β-N-acetylglucosaminidase F	58
Exoglycosidase	57, 58
α-L-Fucosidase	55, 64
α-Galactosidase	51, 53–58, 68
β-Galactosidase	53–56, 58, 59
α-Glucosidase	68
Invertase	84, 86
Lysozyme	56
α-Mannosidase	54–57
Neuramidase	68

[1]Enzymes are listed with the page numbers on which they are mentioned

Glycosyltransferases

N-Acetylglucosaminyltransferase I	38, 40
N-Acetylglucosaminyltransferase II	38, 40
β(1-3) *N*-Acetylglucosaminyltransferase	49
Fucosyltransferase	76
α(1-4)Fucosyltransferase	38, 40
α(1-2)Fucosyltransferase	59
Galactosyltransferase	63, 76
β(1-4)Galactosyltransferase	38, 40, 49, 59, 84, 86, 90, 94
Glucuronyltransferases	79, 80, 88
UDP-glucuronyltransferases	79
Cyclodextrin α1-4 glucosyltransferase	91, 94
Mannosyltransferase	40
Sialyltransferase	51
α(2-6) Sialyltransferase	59, 71, 84, 86
α(2-3) Sialyltransferase	49, 59, 71

Hydrolyases

Inorganic	71, 76
Lipase	24, 28, 32
Phosphatase	71
RNAse	14

Isomerases and Mutases

Enolase	8
Glucose isomerase	24
Phosphoglyceromutase	8
Phosphoglucose isomerase	116
Phosphopentomutase	28
Superoxide dismutase	112
Triosephosphate isomerase	4, 28
UDP-galactose epimerase	94

Kinases

Acetate kinase	8, 112
Adenylate kinase	13, 68, 71, 112, 116
Fucokinase	76
Glucokinase	114
Glycerokinase	4
Guanylate kinase	13, 68, 71

INDEXES

Author Index

Affiliation Index

Subject Index

126

Production: Kurt Schaub
Indexing: Deborah Steiner
Acquisition: A. Maureen Rouhi
Cover design: Lori Seskin-Newman

Printed and bound by Maple Press, York, PA

Paper meets minimum requirements of American National Standard
for Information Sciences—Permanence of Paper for Printed Library
Materials, ANSI Z39.48–1984 ∞

Other ACS Books

Chemical Structure Software for Personal Computers
Edited by Daniel E. Meyer, Wendy A. Warr, and Richard A. Love
ACS Professional Reference Book; 107 pp;
clothbound, ISBN 0–8412–1538–3; paperback, ISBN 0–8412–1539–1

Personal Computers for Scientists: A Byte at a Time
By Glenn I. Ouchi
276 pp; clothbound, ISBN 0–8412–1000–4; paperback, ISBN 0–8412–1001–2

Biotechnology and Materials Science: Chemistry for the Future
Edited by Mary L. Good
160 pp; clothbound, ISBN 0–8412–1472–7; paperback, ISBN 0–8412–1473–5

Polymeric Materials: Chemistry for the Future
By Joseph Alper and Gordon L. Nelson
110 pp; clothbound, ISBN 0–8412–1622–3; paperback, ISBN 0–8412–1613–4

The Language of Biotechnology: A Dictionary of Terms
By John M. Walker and Michael Cox
ACS Professional Reference Book; 256 pp;
clothbound, ISBN 0–8412–1489–1; paperback, ISBN 0–8412–1490–5

Cancer: The Outlaw Cell, Second Edition
Edited by Richard E. LaFond
274 pp; clothbound, ISBN 0–8412–1419–0; paperback, ISBN 0–8412–1420–4

Practical Statistics for the Physical Sciences
By Larry L. Havlicek
ACS Professional Reference Book; 198 pp; clothbound; ISBN 0–8412–1453–0

The Basics of Technical Communicating
By B. Edward Cain
ACS Professional Reference Book; 198 pp;
clothbound, ISBN 0–8412–1451–4; paperback, ISBN 0–8412–1452–2

The ACS Style Guide: A Manual for Authors and Editors
Edited by Janet S. Dodd
264 pp; clothbound, ISBN 0–8412–0917–0; paperback, ISBN 0–8412–0943–X

Chemistry and Crime: From Sherlock Holmes to Today's Courtroom
Edited by Samuel M. Gerber
135 pp; clothbound, ISBN 0–8412–0784–4; paperback, ISBN 0–8412–0785–2

For further information and a free catalog of ACS books, contact:
American Chemical Society
Distribution Office, Department 225
1155 16th Street, NW, Washington, DC 20036
Telephone 800–227–5558